BIOPHYSICS

An Introduction

Biophysics
An Introduction

by
CHRISTIAAN SYBESMA

Kluwer Academic Publishers

DORDRECHT / BOSTON / LONDON

Library of Congress Cataloging in Publication Data

```
Sybesma, C.
    Biophysics, an introduction / by Christiaan Sybesma.
        p.    cm.
    Rev. ed. of: An introduction to biophysics. 1977.
    ISBN 0-7923-0029-7.   ISBN 0-7923-0030-0 (pbk.)
    1. Biophysics.   I. Sybesma, C.   Introduction to biophysics.
    II. Title.
    QH505.S857 1989
    574.1'91--dc19                                        88-30781
                                                              CIP
```

ISBN 0-7923-0029-7 (HB)
ISBN 0-7923-0030-0 (PB)

Published by Kluwer Academic Publishers,
P.O. Box 17, 3300 AA Dordrecht, The Netherlands

Kluwer Academic Publishers incorporates
the publishing programmes of
D. Reidel, Martinus Nijhoff, Dr W. Junk and MTP Press.

Sold and distributed in the U.S.A. and Canada
by Kluwer Academic Publishers,
101 Philip Drive, Norwell, MA 02061, U.S.A.

In all other countries, sold and distributed
by Kluwer Academic Publishers Group,
P.O. Box 322, 3300 AH Dordrecht, The Netherlands.

Printed in The Netherlands

Contents

Preface

Today, courses on biophysics are taught in almost all universities in the world, often in separate biophysics departments or divisions. This reflects the enormous growth of the field, even though the problem of its formal definition remains unsettled. In spite of this lack of definition, biophysics, which can be considered as an amalgamation of the biological and the physical sciences, is recognized as a major scientific activity that has led to spectacular developments in biology. It has increased our knowledge of biological systems to such an extent that even industrial and commercial interests are now beginning to put their stamps on biological research.

A major part of these developments took place during the last two decades. Therefore, an introductory textbook on biophysics that was published a dozen years ago (C. Sybesma, An Introduction to Biophysics, Academic Press, 1977) no longer could fulfil "...the need for a comprehensive but elementary textbook...-" (R. Cammack, Nature **272** (1978), 96). However, because of the increased proliferation of biophysics into higher education, the need for introductory course texts on biophysics is stronger than ever. This fact, together with valuable comments of many readers, have encouraged me to revise the original book.

The basic conception of this new textbook is the same as that of the original book: biophysics is considered as an approach to biology from the conceptual viewpoint of the physical scientist. It is an integrating scientific activity, rather than just an application of physics or physical chemistry to biology. Like in the original book, the major emphasis is on fundamental biological problems and not so much on problems related to medical physics or biotechnology, although, obviously, reference to the latter sometimes could not be avoided.

This new version has twelve chapters. After an introductory chapter, in which the relation between the physical sciences and biology is discussed and an approach to biology based on the concepts of the physical sciences is developed, the universal aspects of the structures of life, the cell and its components, are described in Chapter 2. In this chapter, new insights about cell structure, including those on cytoskeleton, are described and the discussion about membranes is substantially extended. Chapter 3 brings together a description of the physical aspects of (bio)molecular structure and of physical methods used in biophysical (biochemical) research. The reason for treating these two topics together in one chapter is that, after all, the physics of molecular structure lies at the basis of many methods of biophysical measurement. Discussions of techniques now widely in use, such as Raman spectroscopy, chromatography, gel electrophoresis, and magnetic resonance spectroscopy, have been included, as well as a somewhat enlarged and more thorough section on X-ray diffraction analysis. Chapter 4 then discusses molecular and supermolecular structure and function of proteins and nucleic acids; it contains also a short description of recombinant DNA technology. Chapters 5 to 8 describe the energy converting apparatus and membrane transport. A chapter on the biophysics of nerves (Chapter 9) then precedes discussions of contraction (Chapter 10) and of sensory systems (Chapter 11). In the latter chapter a description of the latest developments in our knowledge about signal transmission is included. The final chapter deals with theoretical biology, an area of biology that emerged much as a result of the physical sciences-approach to biology. Its impact on biology may be considerable, not only because it improves our understanding of existing problems but also because it often leads to and determines the direction of experimental biophysical research.

An attempt is made to keep the subject matter of the book strictly on an introductory level without sacrificing too much of the rigor in the treatment. The book is aimed at undergraduate or first year graduate students with a major in physics and/or chemistry, who wish to get acquainted with biophysics in order to be able to determine which area of the field they want to explore in depth. Some knowledge of elementary calculus, quantum mechanics, thermodynamics and some physical chemistry is assumed. If such a background is lacking, readers may find some help in the two appendices on the concepts of quantum mechanics and (equilibrium) thermodynamics.

I already referred to the many comments I received on the previous book, comments which were an important factor in my decision to rewrite it. A number of these comments included valuable suggestions for which I am particularly grateful. To mention all those who were instrumental in the preparation of this text by name would be impossible. I feel, however, that special thanks are due to Mrs. Saskia Vandenbranden-Laame, who painstakingly took care of the wide diversity of illustrations, and to Mrs. Martine De Valck-Van de Perre, who did so much to put the many versions of the text on floppy.

A final word to my readers: my intention in writing this book was to convey the notion that scientific activity, nowadays, can only be successful if it transgresses the boundaries between the disciplines which in the past were so rigidly distinct. For, indeed, biophysics is a scientific activity that does not just overlap with several of the classic disciplines but that integrates them into a science that surpasses those disciplines. If I have conveyed such a notion, I may also have given a deeper meaning to K.S. Cole's somewhat ironic, but good-natured, definition of biophysics: "Biophysics includes everything that is interesting and excludes everything that is not".

C.S.
Brussels, May 1989

1. Introduction

1.1 What is biophysics?

What the definition of biophysics should be is a question asked in almost all texts dealing with biophysics or biophysical sciences. Many of the answers given in such texts are as vague as they are negative: biophysics is a discipline without a fixed content; biophysics is not yet an established discipline; its subject matter is not (yet) very well defined; biophysics is more or less what individual biophysicists have made and are making it. Yet many eminent scientists whose work has significantly contributed to our understanding of many biological phenomena have called themselves biophysicists and apparently did so for more or less pragmatic reasons. This indicates that there is an area of scientific activity which, by general consent rather than by definition, is covered by the name biophysics. Therefore, rather than to try to define or emphasize the lack of definition, it seems to make more sense to identify it by discussing the relation between the physical sciences and biology.

Development of physics

Physics is an exact science which gradually obtained this characteristic in the days following those of Copernicus and Galileo. Its power is its exactness or, in other words, the fact that it deals with accurately measured quantities which allow, often through abstractions, causal interrelations in terms of a sophisticated conceptuality. Many of man's intellectual endeavors were initially descriptive sciences (limited to a compilation of appearances and events) and later developed into more exact sciences. Chemistry ceased to be a descriptive science when scales entered the laboratory. The introduction of quantitative measurements in chemistry resulted in the important laws of conservation of matter (Lavoisier) and constant and multiple proportions (Boyle), followed by Dalton's atom and Mendeleyev's periodic system of the elements. But, while chemistry used to be a science quite distinct from physics, the more it became an exact science, the closer it moved toward physics. Chemistry and physics now appear to be two aspects of the same thing. A dramatic demonstration of this is the reconciliation of Dalton's chemical atom and the physical atom of Rutherford and Bohr. The stability and valence characteristics of Dalton's atom were inexplicable with the Rutherford-Bohr model. The development of quantum mechanics brought about the connection between the two models and initiated an era of scientific activity in which the fundamentals of physics are not different from the fundamentals of chemistry.

Development of biology

The development of biology progressed at a slower pace. This science still exists as a collection of numerous subdisciplines. Many of these, such as botany, zoology, entomology, mycology, and ornithology have long been descriptive sciences and most of them still have that flavor. The development of biology, however, depended on and was connected to that of medicine. Progress in the art of healing is closely linked to the development of disciplines such as anatomy, bacteriology, pharmacology, and physiology. Developments in these areas are responsible for the spectacular advances made in recent times in medicine but also made these areas (especially physiology) more general. A great stimulus for this development was the emergence of biochemistry, after the shattering, by Whöler's synthesis of urea in 1828, of the old notion that organic matter could not arise outside the living organism.

Relation between physics and biology

At that time, physiologists also began to incorporate generalized physical concepts such as mechanics, hydrodynamics, optics, electricity, and thermodynamics into their science. Something like a new subdiscipline began to emerge and many people now see this as the beginnings of biophysics. If we try to trace the history of this penetration of the physical sciences into biology we may as well start with Galvani and his work in the eighteenth century on the effects of static electricity on frog muscle. Julius Robert Mayer, in the nineteenth century, was trained in medicine but is well known for his formulation (in 1842) of the law of conservation of energy as a general principle. He was the first to point out that the process of photosynthesis is essentially an energy conversion process. Hermann Ludwig von Helmholtz (also trained as a physician) is probably the most outstanding example of a successful early biophysicist because of his contributions which led to a greater understanding of both biological and physical concepts. He studied muscle contraction, nerve impulse conduction, vision, and hearing, developed instrumentation to analyze the frequencies of speech and music, invented ophtalmometry, and contributed to thermodynamics. A historical example of a trained physicist who contributed to biology is John Tyndall, who studied under Faraday. His contributions to microbiology as an English contemporary of Louis Pasteur are well known.

The incorporations of physical concepts and instrumentation into biology became more firmly established in this century. The development of the technique of structure analysis by X-ray diffraction had a profound influence on biology. Thermodynamics became an essential part of the study of muscle contraction, and the recent development of non-equilibrium thermodynamics has its direct implications in biology. Electrophysiology obviously cannot be thought of without the concepts and the technology of electricity and electronics. Spectroscopy is an indispensible tool in many areas of physiology, and quantum mechanics is the basis for the interpretation of structure-function relationships at the subcellular and molecular levels in biological systems. The direct implications for biology of areas such as information theory, computer science, artificial intelligence, and cybernetics are promising and

2

further exploration of their applicability may well lead to new levels of understanding in biology.

This kind of biophysics thus can be considered an offspring of physiology. Its feedback to physiology has helped separate physiology from medicine and make it a science in its own right. Biophysics developed into a major part of biology through incorporation of not only existing physical theories but also the conceptual approaches of the physical scientist. It thus became an interdisciplinary scientific activity which surpasses the original organism-defined departments of biology. It is beginning to play the role of an *integrating* science, putting together several specialized branches of science into one context. One of the most important achievements of biophysics is the recognition that the laws of physics and chemistry are valid in living systems; there is no longer any doubt that the fundamental basis of the processes in a living organism is not different from the fundamental basis of the processes of physics and chemistry.

1.2 The fundamental principles of biology

Living versus nonliving
One can, however, also identify another, more recent origin of biophysics which did not emerge from (medical) physiology. It came from physicists and physical chemists who, in their search for universal principles that will explain the world around us, are impressed by the strange and very special place occupied by living organisms. Biophysics viewed from this attitude truly concerns the fundamental principles of biology.

Almost since man was able to think, he wondered about what makes him and the living world around him distinct from the rest of his environment. Scientists as well as philosophers have developed theories about it, some of them based on religious presuppositions, others devoid of such presuppositions. Until well into the nineteenth century, virtually all these theories were of a vitalistic or animistic nature; in other words, they assumed special, not necessarily physicochemical, "forces" to explain living systems as distinct from the nonliving. Even today such vitalistic or animistic theories, either disguised or not, have strongholds. But in the course of the nineteenth century a mechanistic view of life began to develop which states that indeed physics and chemistry are all that is necessary to explain the phenomenon of life.

A biological complementarity principle
The more that was known about the molecular processes of living systems, the more this mechanistic view developed. For many eminent scientists, however, it proved difficult to accept a theory based upon known physical principles that would account for the obvious differences between the living and nonliving world. Evolution over a long period of time leading from a relatively simple self-reproducing system to the wide diversity and complexity

3

of the living world of today seemed to defy the second law of thermodynamics; also, the faithful replication of species over numerous generations has puzzled many of those who were looking for universal explanations. In 1949 Max Delbrück, a physicist, gave a lecture in which he discussed biology from a physical science point of view. Reflecting on the tremendous enrichment process of evolution, he concluded that the principles on which an organism of today is based must have been determined by a couple of billion years of evolutionary history. He then said "...you cannot expect to explain so wise an old bird in a few simple words." Niels Bohr, one of the founders of modern physics, was also impressed by what he considered as the singularity of living organisms. He saw in biology an uncertainty principle analogous to that of quantum theory, but at a higher level; to find out about life in a living organism one has to make measurements which interfere so strongly with the processes in the organism that life itself is destroyed. In 1933 Bohr proposed a formalization of this dualism in the form of a complementarity principle of biology. Much later still a similar opinion was expressed by Werner Heisenberg, a founder of quantum mechanics, who stated in 1962 that "It may well be that a description of the living organism that could be called complete from the standpoint of the physicist cannot be given, since it would require experiments that interfere too strongly with the biological functions." These ideas are the basis of a biological theory developed by Walter Elsasser, a theoretical physicist (1958). We shall briefly discuss his theory in Chapter 12.

Reductionism in biology
In the years before the second world war, however, physical methods rapidly gained acceptance in biological research and many physical scientists began to show an avid interest in their application. Then in 1943 Erwin Schrödinger, one of the founders of quantum wave theory, gave four lectures in Dublin on the physical aspects of a living cell. Later (1944) these lectures appeared in the book "What is Life?" What is remarkable about the book is that slightly more than a decade before the actual discovery of the structure of the gene, the DNA molecule, and the genetic code enclosed in the linear sequence of the nucleotides, these exact ideas were developed. Schrödinger did not start from any fundamental dualism other than that of quantum mechanics itself. He saw the necessity for what he called "aperiodic crystals" as the governing agents of the hereditary "codescript" in the gene. Noticing the smallness of the gene and the tremendous amount of information it must contain to cause the development of a complex living organism, he stated that "A well-ordered association of atoms, endowed with sufficient resistivity to keep its order permanently appears to be the only conceivable material structure, that offers a variety of possible arrangements sufficiently large to embody a complicated system of determinations within a small spatial boundary" [p. 65]. Since then the progress of molecular biology, based on pure physical and physicochemical principles, has been quite impressive.

These events are seen by many as the beginning of an era in which physical science and physical concepts were being unconditionally accepted in the study

4

of biology. Since then, biological research developed in a full-scale application of reductionistic methods. These methods have been extremely successful in the physical sciences and certainly account for the spectacular advances made in biology since World War II. Reductionism tries to explain the behavior of a composite system by the properties and behavior of the parts of the system. Indeed, it has led to a better understanding but especially to a better control of nature, including life. We all recognize the technological achievements, also in biology, of our time. However, somewhat paradoxically it also led to new insights, coming from recent developments in the physical sciences, that reductionism has limits in its explaining power. These insights may exert a profound influence on our understanding of life and its place in the cosmos.

In the following pages of this book we introduce biophysics as an approach to biology that is characterized not only by the description of the physico-chemical basis of the processes in living systems, but also by a physical conceptuality of the subject matter. We shall first see why and how a living system can be generalized and then discuss how physical theories and concepts apply to such generalized systems by looking at diverse interactions and processes occurring in these systems, from the molecular level up to more integrated systems. In the final chapter these generalizations will reappear in a discussion of theoretical biology and we shall try to find out what they mean with respect to questions about the fundamental principles of biology.

Bibliography

Bohr, N. (1933). Light and Life, *Nature* **131**, 421–423.
Delbrück, M. (1949). A Physicist Looks at Biology, *Trans. Conn. Acad. Arts Sci.* **38**, 175–190.
Elsasser, W.M. (1958). "The Physical Foundation of Biology." Pergamon Press, Oxford.
Heisenberg, W. (1962). "Physics and Philosophy." Harper & Row, New York.
Schrödinger, E. (1944). "What is Life?" Cambridge Univ. Press, London.
Stent, G. (1968). That was the Molecular Biology that was, *Science* **160**, 390–396.

2. Biological structures

2.1 The structures of life

The cellular structure of life
In the beginning of the twentieth century, the study of matter by both physicists and chemists was simplified and unified by the final description, in quantum mechanical terms, of its supposedly fundamental unit, the atom. Although the atom presently is known not to be a fundamental but a composite particle, one can still consider the concept as basic for the description of matter. The properties of all substances can be explained in terms of the physics of this basic unit.

Does such a concept also exist in biology? In many biology textbooks one finds the proposition that the living cell is such a concept. The emergence of the cell doctrine after Robert Hooke's discovery of the "little boxes or cells" in cork in 1665, resulted in the general recognition that life has a cellular structure. In spite of the wide variety of size, shape, and function, the cell, as an integral and relatively independent body surrounded by a boundary, is a common feature of all living organisms and seems therefore very suitable to be a generalizing concept in biology. The cell, indeed, appeared to be an essential life structure; when a cell is broken the fragments, after some time, lose the characteristics they had and the reactions they were able to carry out in the intact cell, although the material components are not changed.*

Of course, a cell consists of many substances, small and large molecules, complexes, aggregates and so on, all of them being constituted by atoms. Thus, as far as matter can be described by using the concept of the atom, the latter is a basic element of life as well. But a cell is more than just a composite of substances. A haphazard heap of all these substances is not living. Apparently, there is something that transcends this basic element.

Organization
That "something" may be called organization, or better perhaps, organized complexity. Organization is found on all levels of life. Life, indeed, consists of a hierarchy of organizational forms, ranging from society via man and multicellular organisms to the unicellular organisms, protozoa, algae, and

---

* A virus has no cellular structure. A virus, however, cannot be considered as an independent living organism. It functions only *within* a host cell, which it requires to reproduce itself. Outside a living cell a virus behaves like, and can be considered as, lifeless matter.

bacteria. If we upset this organization from the top of the hierarchical ladder on down we end up at the cell. Societies can be broken apart, after which new ones are formed (history is full of examples of such events); organs of multicellular organisms can be removed from the organism and kept alive for an indefinite time under the proper conditions. Organs can be dissociated into their component cells and such cells can be cultivated (tissue cultures) and kept functioning for countless generations; even cells dissociated from well-differentiated livers or kidneys from chicken embryos have been shown to develop miniature livers or kidneys. But when a cell is broken apart the components do not develop into a new cell; they can perform specific functions such as fermentation, respiration, primary photosynthetic processes, or even synthesis of the macromolecules which life is making use of but they die off after some time. The cell, therefore, seems to be the lowest form of organization of matter which we can call life.

Cells occur in life in a wide variety, not only of sizes (a nerve cell can be longer than a meter while a pneumococcus is about 0.2 μm in diameter), but of shapes and functions as well. Figure 2.1 shows some shapes of cells from multi- and unicellular organisms. They all have in common, however, a high-precision apparatus for energy-providing, synthesizing, and specific functional reactions, a center for the control of these reactions, and pathways of communication. This machinery is built up along surprisingly uniform lines; the same type of molecules and the same types of structural features are associated with these functions. The all-important synthesis of proteins, for example, takes place according to the same metabolic patterns regardless of the biological origin of the cell. Thus, it seems as if, from a biophysical

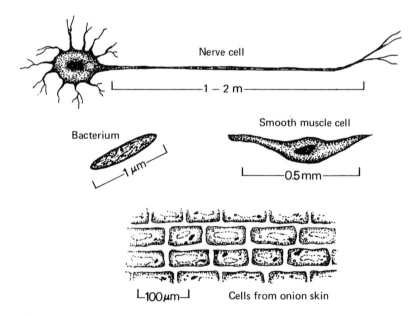

Fig. 2.1. Size and shape of some cells.

8

(biochemical) point of view, we can allow ourselves this generalization in biology and consider the structural and functional features of cells as fundamental for our study of life.

Membranes

A common structural feature in all cells is the membrane. Membranes are found both as boundaries of the cells as well as more or less elaborate systems within the cell. Many biochemical events take place on or in the membranes. Formerly considered as a static structure with a principal role of simple delineation of the boundaries of cells and subcellular organelles, these membranes are now looked upon as highly dynamic structures essential for many diverse processes. In section 2.4 we will discuss the most important features of membranes.

The membrane surrounding the cell is called the *cytoplasmic membrane* (or *plasmalemma* or *plasma membrane*). It encloses the *cytoplasm*, which is the inner part of the cell, and forms a selective barrier which maintains the chemical integretity of the cell. This is done, as will be discussed later, by passive and active transport processes across the membrane which are selective not only in the rate of the movement in or out but also in the kind of molecules that enter or leave the cell. There are other ways for the intake of matter

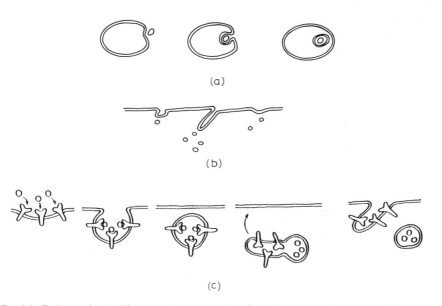

(a)

(b)

(c)

Fig. 2.2. Endocytosis. (*a*) *Phagocytosis*; a protrusion of cytoplasm comes in contact with a drop of liquid or a solid particle, surrounds it, and draws it into the cytoplasm where it can be digested. (*b*) *Pinocytosis*; the cytoplasmic membrane forms channels through which liquid can flow into the cell. In the cell the channels are pinched off as membrane-enclosed droplets. These droplets eventually dissolve in the interior of the cell. (*c*) *Receptor-mediated endocytosis*; a specific ligand binds to a specific receptor binding site provided by a membrane-bound protein, triggering an invagination of the membrane. The membrane pinches off and is digested in the cell thus bringing the ligand into the cell.

9

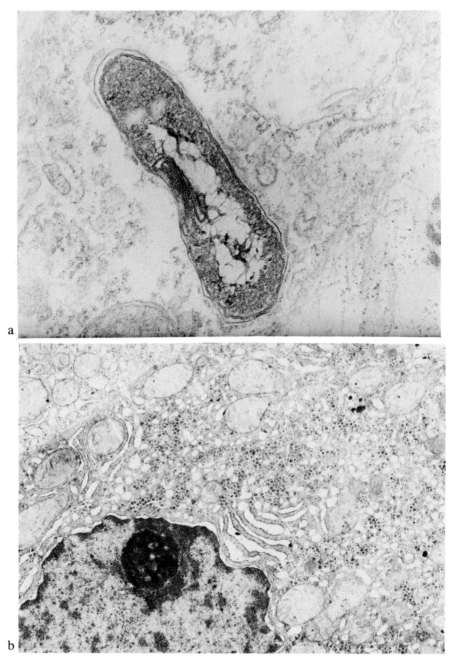

Fig. 2.3. (*a*) Electron micrograph of a prokaryotic organism *Myobacterium leprae*. The cytoplasmic membrane convolutes sometimes in vesicle-like structures and sometimes into stacks. The nuclear area is not well defined. (*b*) Electron micrograph of part of a human liver cell. The nucleus is a well defined area bounded by the nuclear membrane. The black round spot in the nucleus is a nucleolus. The circular structures in the cytoplasm are the mitochondria. The endoplasmic reticulum is also visible as are many ribosomes. Part of the cytoplasmic membrane is seen in the upper left-hand corner. Courtesy of Dr. W. Jacob and Dr. F. Lakiere, Electron Microscopy Laboratory, Universitaire Instellingen Antwerpen, Antwerp, Belgium.

into cells however. Especially large molecules and particulate matter can be brought into a cell by a process called *endocytosis* (Fig. 2.2). There are three kinds of endocytosis. In *phagocytosis* (Fig. 2.2a) a large particle or a molecular complex makes contact with the cell surface and triggers an expansion of the membrane. This expansion engulfs the object, which is then drawn into the cytoplasm where it can be digested. *Pinocytosis* (Fig. 2.2b) is endocytosis that results in the nonspecific intake of fluid from the cell's environment. The cytoplasmic membrane folds inward, forming channels leading into the cytoplasm. Liquids can flow into these channels and the membrane then pinches off pockets that are incorporated into the cytoplasm and digested.

The third kind of endocytosis is very specific. It is called *receptor-mediated endocytosis*. This is endocytosis in which receptor-proteins incorporated in the membrane bind specifically to a ligand (a protein molecule or a small particle) that fits the binding site of the protein (Fig. 2.2c). An invagination of the membrane occurs which forms a membrane vesicle containing the ligand. The membrane pinches off and the vesicle is drawn into the cytoplasm where it is digested, thus bringing the ligand into the cytoplasm.

The cytoplasmic membrane of a cell can thus be seen as a functional part of the living cell. This view is substantiated by the fact that the internal membrane systems in some cells appear to be continuations of the cytoplasmic membrane (see, for instance, Fig. 2.3a). At the same time, there must be a wide variety in the detailed molecular composition of membranes in order to account for the variations in permeability, functions, and appearances. This variability has its origin in the great variety of proteins and lipids that makes up the structure of membranes (see section 2.4). Membranes indeed are a structural feature essential for life processes.

2.2 The morphology of cells

Cell types

Presently, a major morphological distinction between groups of cells is generally accepted. The majority of cells belong to one group, the so-called *eukaryotes*. Bacteria, including cyanobacteria (formerly called blue-green algae) belong to another group, the so-called *prokaryotes*. This nomenclature, proposed by E.C. Dougherty in 1957, is derived from the Greek word καρυον which means kernel or nucleus. The prokaryon (πρo = before) is the undeveloped nuclear area in the prokaryotes and the eukaryon (ευ = well) is the well-developed nucleus in the eukaryotes. Figure 2.3a shows a cross section of a prokaryotic cell, and Fig. 2.3b shows a cross section of a part of a eukaryotic cell. While in the prokaryotic cell no well-developed internal structure in the form of more or less extensive particulate organelles can be seen, the eukaryotic cell shows a highly differentiated internal structure, with visible particulate organelles. This type of cell has a well-defined nucleus surrounded by nuclear membrane. In prokaryotic cells a well-defined nucleus cannot be distinguished,

11

although areas with higher concentration of nuclear material than the rest of the cell can sometimes be recognized. Prokaryotic cells, instead of having systems of particulate organelles, have more or less elaborate membrane systems showing up as either vesicle-like structures or irregular stacks (Fig. 2.3a).

Recently, a third group of cells has been defined that have properties different from those of both eukaryotes and prokaryotes. These are the *archaebacteria* which apparently came into existence very early in evolution (see section 2.5). The most pronounced difference with the other two groups of cells is in the structure and composition of their membranes. The methane bacteria belong to this group.

Cytoskeleton

The cytoplasm of a (eukaryotic) cell is not an amorphous fluid mass in which the nucleus and other organelles are randomly scattered. On the contrary, it is highly structured. Spanning the cytoplasm between the nucleus and the inner surface of the cytoplasmic membrane is a fibrous matrix of proteins that seems to be fundamental in establishing the cell's shape and to play a role in cell locomotion and division. This system of protein fibers is known as the *cytoskeleton* (see Fig. 2.4).

Presently one recognizes three groups of fibers. The finest of the fibers are the microfilaments (on the average 6 nm in diameter). They are made up of a protein called *actin* (see also section 10.3). Microfilaments seem to be crucial to cell locomotion and surface movements. If an immunological technique is used, an array of long, thick bundles of microfilaments, the *stress fibers*, can be seen running parallel to and just inside the cytoplasmic membrane (Fig. 2.4a). Stress fibers play a role in stabilizing the shape of the cell.

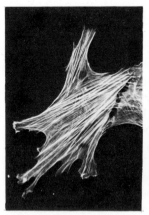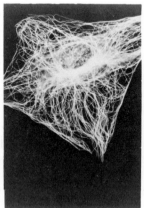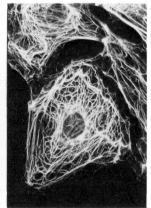

Fig. 2.4. Cytoskeleton of some cells; (*a*) stress fiber arrangement in a rodent cell line, revealed with antibody to actin; (*b*) microtubular arrangement in a rodent fibroblast, revealed with antibody to tubulin; (*c*) intermediate filaments, revealed with antibody to keratin. Magnification 900×. Note the different arrangements of the different cytoskeletal elements. Courtesy of Drs. M. Osborn and K. Weber, Max-Planck-Institute for Biophysical Chemistry, Göttingen, Federal Republic of Germany.

12

By the same kind of technique, another group of fibers, about 22 nm in diameter, can be made visible. These fibers are the *microtubules* (Fig. 2.4b). They consist of another protein called *tubulin*, and radiate from a sort of organizing center near the nucleus, known as the *centrosome* (or *cytocenter*). The centrosome consists of two compact cylindrical structures some 300-500 nm long and 150 nm diameter called *centrioles*, and the surrounding material. Individual microtubules run from the centrioles to just under the cytoplasmic membrane. At the onset of mitosis, the process of cell division, the microtubular complex rearranges itself. The microtubules break down and the tubulin reassembles into a framework forming the *spindle*, a system of parallel microtubular filaments that extend from the poles of the dividing cell to the two sets of *chromatids* which together constitute the *chromosomes*, the structures that contain the cell's genetic material (see below). The spindle guides and probably supplies the driving force for the movement of the sets of chromatides to the poles after which the cells divides. The two daughter cells thus receive one set of chromatides each. Microtubules appear to establish the geometry of the cell, acting as tracks that orient also other cell phenomena.

Less is known about the third group of fibers, the so-called *intermediate filaments* (Fig. 2.4c). Their protein composition varies according to the type of cell. The arrangement of the fibers is different in different tissues and some cells seem to lack them altogether. In many kinds of cells they are related somehow to the microtubules.

Eukaryotic organelles

Figure 2.5 shows a schematic representation of two types of eukaryotic cells, an "animal" cell (also representative of many eukaryotic protozoa) and a plant cell. Such representations are useful to point out the common features found in real cells. Animal as well as plant cells are surrounded by a cytoplasmic membrane. In addition plant cells are surrounded by a rigid and sturdy cell wall made of cellulose or cellulose-like material (formed by repeating units of glucose). Animal cells do not usually have this feature; their boundary is formed by the cytoplasmic membrane itself. Another structural feature, more characteristic of plant than animal cells, is the large liquid-filled cavity in the middle of the cell called the *vacuole*. The vacuole is part of a transport system and is surrounded by a membrane called the *tonoplast*.

All eukaryotic cells have a well-defined nucleus surrounded by a double-layered membrane, the nuclear membrane. Like the cytoplasmic membrane the nuclear membrane often shows continuity with membrane systems in the cytoplasm. The nuclear membrane has pores, that is small openings at the edges of which the inner and outer membrane are continuous. The material inside the nucleus is a very fine thread-like substance called *chromatin*. The chromatin consists of the nucleic acids deoxyribonucleic acid (DNA) and ribonucleic acid (RNA), low molecular weight proteins called *histones*, and some residual protein. A nucleus may contain one or more denser bodies called *nucleoli* (singular: *nucleolus*).

The nucleus is an essential structure of eukaryotic cells. The DNA contains

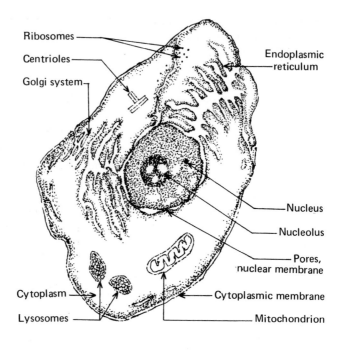

Ribosomes

Centrioles

Golgi system

Endoplasmic reticulum

Nucleus

Nucleolus

Pores, nuclear membrane

Cytoplasm

Lysosomes

Cytoplasmic membrane

Mitochondrion

(a)

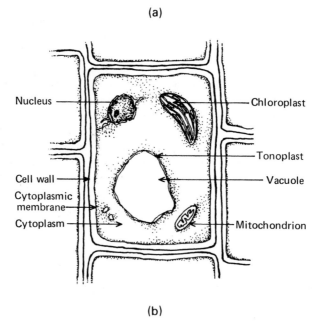

Nucleus

Chloroplast

Tonoplast

Cell wall

Vacuole

Cytoplasmic membrane

Cytoplasm

Mitochondrion

(b)

Fig. 2.5. Schematic representations of (*a*) a typical "animal cell" and (*b*) a typical "plant cell". The figures describe the cells and the organelles in a schematic way, not as they would appear in reality. Animal cells are bounded by just the cytoplasmic membrane; most plant cells have an additional rigid and sturdy cell wall around them.

14

all the information for the morphology and function of the cell. When the cell divides, the chromatin becomes visible as elongated structures, the chromosomes that we have already mentioned. The chromosomes are made up of pairs of chromatids which, guided by the spindle, are separated (see above). Each chromatid forms an equal set of genetic material for the two daughter cells. The DNA of the chromatin is exactly duplicated by a process which will be described in Chapter 4.

Both types of eukaryotic cells have more or less elaborate membrane structures in their cytoplasm. These channel- or sac-forming (cisternal) structures, known under the name *endoplasmic reticulum*, extend to a variable degree from the nuclear membrane to the cytoplasmic membrane. The type of cell and its state of metabolic activity can be characterized by the appearance of the endoplasmic reticulum. This structure seems to have a variety of functions, an important one being, most probably, that of providing a communication system in the cell; thus, products formed can be segregated and transported either to other parts of the cell or to the outside environment by this system of channels. Surrounding part of it and sometimes attached to it are numerous little granules called *ribosomes*, giving this part of the endoplasmic reticulum a rough appearance. Ribosomes, as we will see later, are instrumental for the synthesis of proteins.

Another set of membranes seen in both types of cells is the *Golgi apparatus*, sometimes called the *dictyosome*. This peculiar membrane system acts as a packaging and transport system for products and waste by *exocytosis*, a process which can be seen as a reversed endocytosis. Specialized cells with a secretory function have a well-developed Golgi system. In such cells products synthesized at or near the endoplasmic reticulum are packaged within membrane-enclosed granules which arise from the Golgi apparatus. These vesicles subsequently migrate toward the cell surface, fuse with the cytoplasmic membrane, and discharge their content into the intercellular space by exocytosis. Such a function may be indicated by, among other things, the appearance of the system; deep inside the cell the vesicles are elongated and flattened and arrayed in a more or less regular parallel stack close to the endoplasmic reticulum. Toward the edge of the cell the vesicles become less flattened and often develop into a large number of irregular but more or less rounded vesicles. The system is not a static one; there seems to be a constant flow of membranes from the inside to the outside of the cell, the replenishment originating from the endoplasmic reticulum.

Lysosomes are vesicle-type organelles with an outer limiting membrane. They contain so-called *lytic* enzymes involved in the breakdown of cellular fragments and large molecules. Lysosomes, therefore, can be considered as disposal units of the cell, removing foreign bodies and cell structures no longer needed. There is strong evidence that the lysosomes originate in the Golgi apparatus. This would be consistent with the functions ascribed to both organelles.

The above-mentioned ribosomes, little granules of about 20-25 nm diameter, are found in every kind of cell. They are intimately connected with the synthesis of proteins, consist of two parts, and contain ribonucleic acid (RNA) and protein. Ribosomes originate in the nucleolus but are assembled in the cytoplasm. Aggregates of them are called *polysomes* and are formed in the process of translation which will be described later (section 4.4).

Every kind of eukaryotic cell has at least one, but normally more, *mitochondria* (singular: *mitochondrion*). This extremely important, and probably most studied, organelle is the "power plant" of the cell. Mitochondria range in size from 0.2 tot 7.0 μm, vary in shape from spheres to more or less elongated rods, and are surrounded by a smooth membrane wall. On the inside, some 6 nm from the outer membrane, an inner membrane convolutes inward (in the so-called *matrix*) forming thin sheets called *cristae* (see Fig. 2.6). On these cristae are found the enzymes which operate in the conversion of energy obtained in the aerobic part of the breakdown of carbohydrates, fats, and (to a certain extent) proteins, in a process called *respiration*. This process will be described in more detail in Chapter 7.

Cells found in the green parts of plants, especially in the leaves, and in most

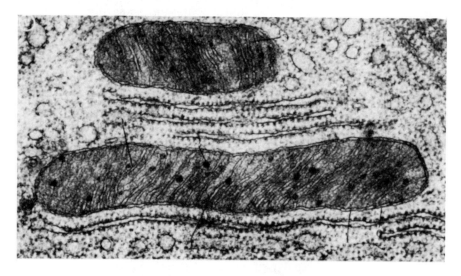

Fig. 2.6. (*a*) Electron micrograph of mitochondria from a pancreas cell of a bat and (*b*) a drawing of a bisected mitochondrion. The electron micrograph clearly shows the many folds of the inner membrane sticking out to the inside (cristae), thus forming the lamella shown in the drawing. Courtesy of Dr. D.W. Fawcett.

of the unicellular algae contain *chloroplasts*, organelles which are the site of the phototrophic energy conversion. The process in which light energy is absorbed, trapped, and converted into chemical energy is called *photosynthesis*. It will be described in Chapter 6. The apparatus for this process is bound to a lamellar system inside the chloroplast, which is again enclosed by a membrane. In many chloroplasts, such as the one shown in Fig. 2.7, the lamellar structure is densely packed at some places (the grana) and single-layered at others (the intergranal lamella). The intergranal space is often referred to as the *stroma*.

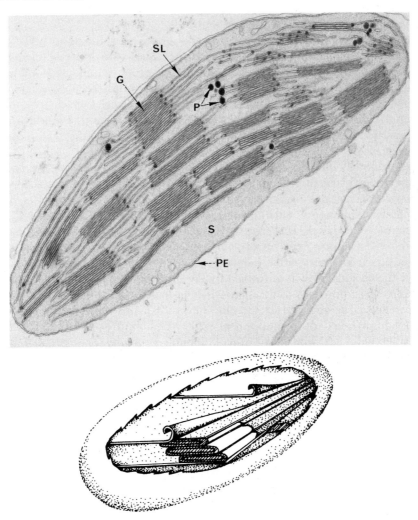

Fig. 2.7. (a) Electron micrograph of a chloroplast from a lettuce leaf cell and a three-dimensional drawing of a cut-open chloroplast. The electron micrograph clearly shows the grana (G) formed by the stacks of flattened vesicles (thylakoids) shown in the drawing (*b*). PE is the chloroplast envelope, a double membrane surrounding the chloroplast. SL are the intergranal lamella, S is the stroma, and P are densely stained lipid droplets (courtesy of Dr. C.J. Arntzen).

2.3 The biological macromolecules

Many different molecules participate in biological reactions and these reactions, in turn, are of a wide diversity. Moreover, this diversity spreads out over the multitude of different kinds of cells and different stages of development. This diversity, however, is not a random one; on the contrary, it is a manifestation of a highly precise system of regulation and control. The system makes use of features built and organized along surprisingly unified lines. The biological macromolecules are the essential part of these features.

Biopolymers
There are four kinds of biological macromolecules. Three of these are polymers, that is long sequences of relatively small molecules (which are of the same kind but not necessarily identical). These are the proteins (polymers of amino acids), the nucleic acids (polymers of nucleotides), and the polysaccharides (polymers of sugars). The fourth group consists of molecules which are known as lipids, loosely defined as that portion of animal or plant tissue which can be extracted with the so-called "fat solvents" such as ethanol, ether, chloroform, and benzene.

Many of the functions of the macromolecules in living systems are intimately bound to their structure. The structures are determined by interactions between the components making up the macromolecules and by interactions of such components with the environment. We can distinguish four "levels" of structure: the *primary structure* is the sequential order of the components (such as the amino acids in proteins, the nucleotides in nucleic acids and the sugars in polysaccharides). The *secondary structure* is the locally ordered three-dimensional structure formed by the succession of components. Linear (head to tail) polymers with asymmetric monomeric units (such as proteins and nucleic acids) organize in screw-like structures called *helices*. The secondary structure of these biopolymers thus is the helical structure. Often, these helical structures can fold into more or less globular structures or join together in fiber-like structures, thus forming the *tertiary structure*. The tertiary structure of a protein or a nucleic acid is the complete three-dimensional structure of one indivisible unit. The *quaternary structure* is the (mostly noncovalent) association of such indivisible units into larger units. The subunits of the quaternary structure may or may not be identical and their arrangement may or may not be symmetric. The levels of structure and the relation between structure and function of proteins and nucleic acids will be discussed in more detail in Chapter 4. Here, we will look at the macromolecular composition (the primary structures) only.

Proteins
A large part of the dry weight of a cell is protein. It is a major class of compounds found in all living matter. Enzymes are proteins which catalyze and control the rates of many biological reactions; muscular contraction depends on the proteins myosin, actin and some others; active transport across

cell membranes and energy conservation seem to depend on the properties of the protein-lipid complexes in the membranes. Furthermore, there are structural proteins which form fibers, sheets and tubules.

The building blocks of proteins are the *amino acids*. These are relatively small molecules consisting of a carboxyl group (I) and a basic amino group (II) attached to a central carbon atom, the so-called α-carbon; the general form of an α-amino acid, thus is (III):

$$
\begin{array}{ccc}
\overset{\displaystyle O}{\underset{\displaystyle OH}{-C{\Large\diagdown}}} &
\overset{\displaystyle H}{\underset{\displaystyle H}{-N{\Large\diagup}}} &
\overset{\displaystyle H\quad H\quad O}{\underset{\displaystyle H\quad R\quad OH}{N-C_\alpha-C}}
\end{array}
$$

$$\quad\text{(I)}\qquad\qquad\text{(II)}\qquad\qquad\text{(III)}$$

Also attached to the α-carbon is a molecular group R, called the *residue*. The representation is a projection in a plane through the α-carbon. In reality the hydrogen and the amino group are above the plane of the paper whereas the carboxyl group and the residue are below it.

Amino acids react with each other to eliminate, between each two of them, a molecule of water, thus forming a peptide bond:

$$
\underset{\substack{H\quad R_1}}{\overset{\substack{H\quad H\quad O}}{N-C-C}}\;[OH \;+\; \underset{\substack{H]\quad H\quad OH}}{\overset{\substack{H\quad R_2\quad O}}{N-C-C}} \longrightarrow \underset{\substack{H\quad R_1\qquad\quad H\quad OH}}{\overset{\substack{H\quad H\quad O\quad H\quad R_2\quad O}}{N-C-C-N-C-C}} + H_2O
$$

These peptide bonds are covalent bonds and very stable. Long chains of amino acids bound together by peptide bonds are called *polypeptide* chains. When 50 or more amino acids are bound this way, we refer to the chain as a protein.

The specific residues present determine the individual amino acids. Many residues exist and accordingly many α-amino acids can be, and have been, synthesized. In the proteins of living cells, however, only 20 of them occur. For proteins or peptides synthesized on the ribosomes (see Chapter 4) all amino acids are L-stereoisomers (showing left-turning optical activity). The residues of these 20 amino acids together with their names and their indicating (one-letter and three-letter) abbreviations are given in Table 2.1.

It seems obvious that the three-dimensional structure of a protein is determined by the mutual interactions of the residues which form side chains sticking out of the protein backbone. These interactions depend largely on the polarity of the residues. Nonpolar residues have no electrical charge and are insoluble in polar solvents such as water. In fact, in their tendency to avoid water the nonpolar chains tend to aggregate with each other (see below). This is particularly true for the aliphatic groups of alanine, valine, leucine and isoleucine which are not restrained by steric hinderances. But also the aromatic residues of phenylanaline, tyrosine and tryptophan behave as nonpolar residues and glycine, which really does not have a side chain, can be seen also as

Table 2.1

Aminoacid	Residu Structure	Abbreviation	
		one letter	three letters
Non polar			
Glycine	—H	G	gly
Proline		P	pro
Aliphatic			
Alanine	—CH_3	A	ala
Valine	—$CH \big\langle\, {}^{CH_3}_{CH_3}$ (with H)	V	val
Leucine	—CH_2—$CH \big\langle\, {}^{CH_3}_{CH_3}$ (with H)	L	leu
Isoleucine	—$CH \big\langle\, {}^{CH_2-CH_3}_{CH_3}$ (with H)	I	ile
Aromatic			
Phenylalanine *	—CH_2— (phenyl ring)	F	phe
Tyrosine *	—CH_2— (phenol ring) —OH	Y	tyr
Tryptophan *	(indole ring structure)	W	trp
Histidine	(imidazole ring structure)	H	his

* Each corner (*C*-atom) of the ring structure has a bound hydrogen.

Table 2.1 Continued.

	Structure	Code	Abbr.
Sulfur containing			
Methionine	$-C_{H_2}-C_{H_2}-S-CH_3$	M	met
Cysteine	$-C_{H_2}-SH$	C	cys
Polar			
Asparagine	$-C_{H_2}-C(=O)-NH_2$	N	asn
Glutamine	$-C_{H_2}-C_{H_2}-C(=O)-NH_2$	Q	gln
Serine	$-C_{H_2}-OH$	S	ser
Threonine	$-C\langle\begin{smallmatrix}H\\OH\\CH_3\end{smallmatrix}$	T	thr
Net positive			
Lysine	$-C_{H_2}-C_{H_2}-C_{H_2}-C_{H_2}-NH_3^+$	K	lys
Arginine	$-C_{H_2}-C_{H_2}-C_{H_2}-N_H-C\langle\begin{smallmatrix}NH_2\\NH_3^+\end{smallmatrix}$	R	arg
Net negative			
Aspartate	$-C_{H_2}-C\langle\begin{smallmatrix}O\\O^-\end{smallmatrix}$	D	asp
Glutamate	$-C_{H_2}-C_{H_2}-C\langle\begin{smallmatrix}O\\O^-\end{smallmatrix}$	E	glu

a nonpolar amino acid. Clearly polar residues are those of asparagine and glutamine (because of their dipole-possessing amide groups), serine and threonine (because of their end hydroxyl groups which also posses strong dipole moments), lysine and arginine (which have a net positive charge) and aspartic acid and glutamic acid (which have a net negative charge). These groups can interact electrostatically with other polar groups or molecules, thus forming noncovalent bonds. The sulfur containing residues of methionine and cysteine can become charged in some environments; in these circumstances they behave as polar residues. If not charged they certainly are nonpolar.

In an aqueous (water) environment the nonpolar residues, avoiding the water molecules, tend to stay inside the protein while the polar molecules are more on the outside.

One type of residue interaction can lead to another covalent bond in the protein molecule, i.e. the bond between the sulfur atoms in cysteine. The dimer of cysteine, formed by the covalent bond between the sulfur atoms of each after the removal of the hydrogen, is called cystine. This dimer is responsible for the so-called sulfur bridges which either link two polypeptide chains together or make a loop in one polypeptide chain (Fig. 2.8).

Nucleic acids
Nucleic acids are the biopolymers responsible for the preservation of biological identity. They are bearers and conveyers of all the information regarding the structure and behavior of the living system. The way in which they accomplish this will be discussed in Chapter 4. Their polymeric character, expressed in sequences of monomeric units in a very stable configuration, is an essential feature for this function. Nucleic acids are polymers of a molecular group called a *nucleotide*. A nucleotide consists of a sugar, a base, and a phosphate, and the polymer is formed by covalent disphosphoester bonds

$$-\overset{\overset{\displaystyle H}{|}}{\underset{\underset{\displaystyle H}{|}}{C}}-\overset{\overset{\displaystyle H}{|}}{\underset{\underset{\displaystyle H}{|}}{C}}-O-\overset{\overset{\displaystyle O}{\|}}{\underset{\underset{\displaystyle O}{|}}{P}}-O-C-$$

When the sugar is a five-membered ring (pentose) with the configuration (I) it is called D-ribose and the nucleic acid is ribonucleic acid (RNA). When the sugar is a five-membered ring (pentose) with the configuration (II), it is called D-2-deoxyribose and the nucleic acid is deoxyribonucleic acid (DNA).

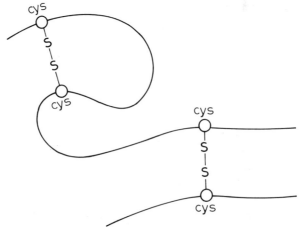

Fig. 2.8. Tertiary structure of protein formed by S bonds.

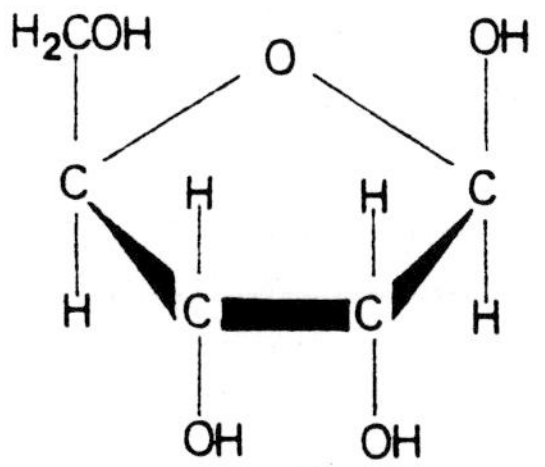

(I) 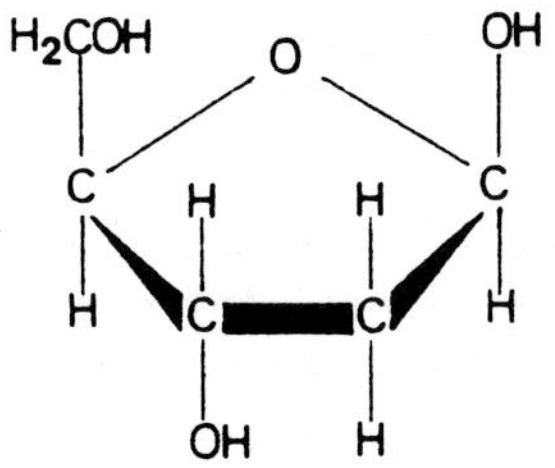   (II)

In both names the D stands for right-turning optical activity. In both formulae R designates the position of the base.

There are two kinds of bases, those derived from *pyrimidine* and those derived from *purine*:*

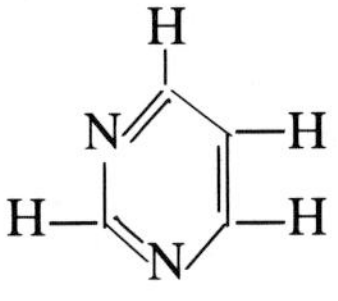

pyrimidine 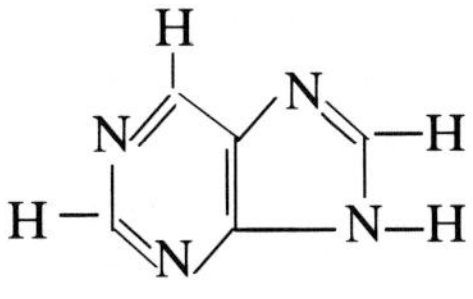   *purine*

The three bases derived from pyrimidine are *cytosine* (C), *thymine* (T), and *uracil* (U):

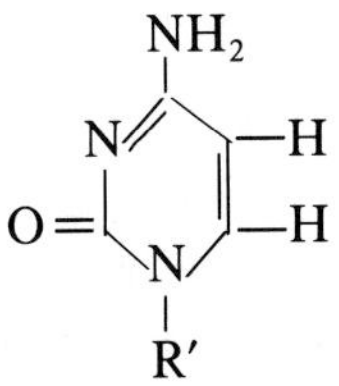

(C) 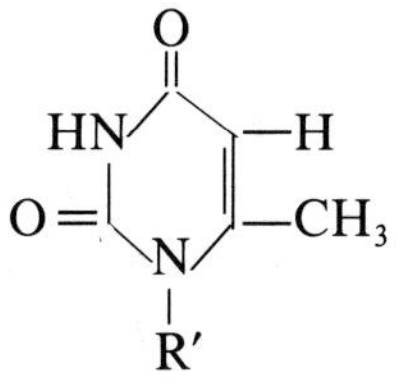   (T) 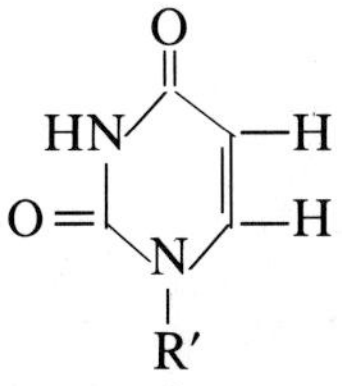   (U)

and the two bases derived from purine are *adenine* (A) and *guanine* (G):

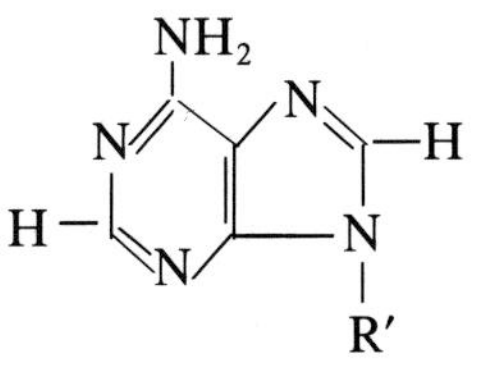

(A) 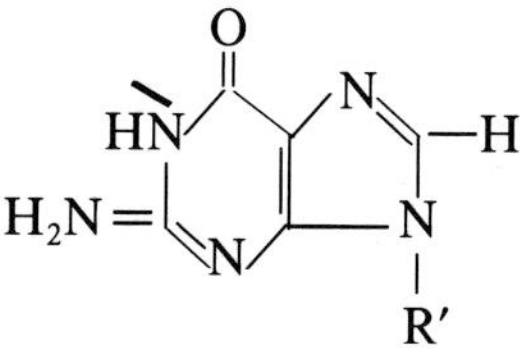   (G)

The R' stands for the sugar to which the base is attached. Each of the two kinds of nucleic acid contains only four of the five bases: In DNA, the pyrimidine-derived bases are cytosine and thymine; in RNA, however, thymine is replaced by uracil.

Figure 2.9 shows the basis chemical structure of DNA. The bases are attached

---

* Usually, the carbon atoms in the corners of ring structures are not indicated.

23

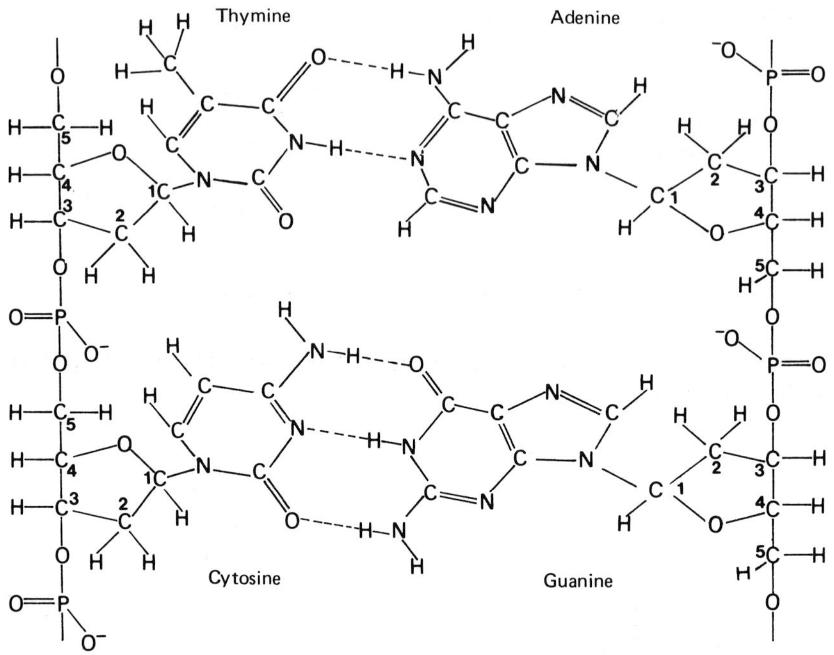

Fig. 2.9. The chemical structure of DNA. The bases thymine, adenine, cytosine, and guanine are attached to deoxyribose phosphate strands. Two such strands are bound to each other by hydrogen bonds between the bases: two between thymine and adenine, three between cytosine and guanine, thus forming the twisted double-stranded structure.

to 1st carbon of the sugars. Most DNA molecules have a double-stranded twisted structure with the bases bound to each other by hydrogen bonds, a type of electrostatic interaction which will be discussed in Chapter 3. The phosphates in the sugar-phosphate sequence are connected at one side to the 5th carbon atom of the sugar and at the other side to the 3rd carbon of the sugar. In respect to the bases, the sequence runs 5–3 in one strand and 3–5 in the other strand. The two strands thus are antiparallel as is shown in Fig. 2.9.

RNA has a single-stranded covalent structure. It occurs in several different types the major ones of which being *messenger* RNA (mRNA), *transfer* RNA (tRNA) and *ribosomal* RNA (rRNA). The structures in relation to the functions of the major types of RNA as well as of DNA will be discussed in more detail in Chapter 4.

Polysaccharides

Polysaccharides are polymers made up of sugars. The most common sugars making up these long chains are hexoses (six-membered carbon rings). An example is amylose, a component of starch that serves as a nutrient reserve in plants. The glucose molecules are in the form of pyranose rings (like ribose and deoxyribose in respectively RNA and DNA) and are linked together

24

covalently by α-glycoside bonds:

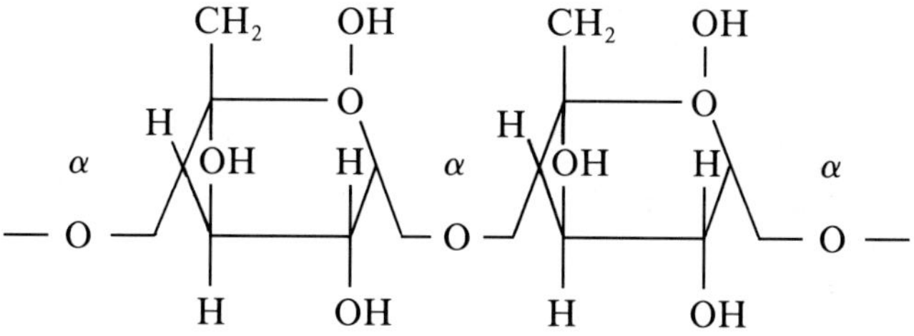

Another example is cellulose, also made up by glucose pyranose rings linked together, however, by β-glycoside bonds:

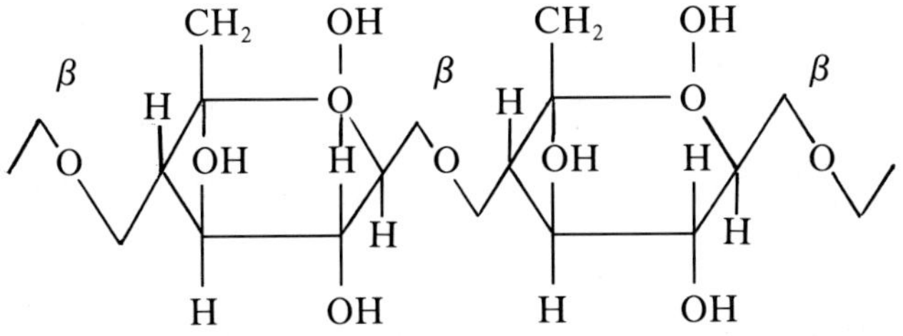

the difference being the stereochemistry at position 1 of the glucose.* Cellulose is a major component of wood and other plant fibers. If the hydroxyl group attached to C-atom 2 in the glucose ring is replaced by

$$-NH-\overset{\overset{\displaystyle O}{\|}}{C}-CH_3$$

we have *chitin* which forms the exoskeleton of insects and other arthropods.

Lipids

The lipids are a heterogeneous group of molecules which can play various roles throughout the cell. They are easily extracted into organic solvents and are hardly soluble in water. As fat (compounds of glycerol and fatty acids) they can store energy which need not be immediately available. A typical neutral fat has the following structure:

---

* The C-atoms in the glucose rings are counted counter-clockwise, starting from the right-side C-atom engaged in the glycoside bond.

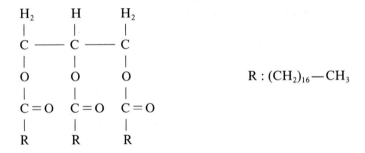

$$R : (CH_2)_{16}-CH_3$$

Such a structure is highly hydrophobic which means practically that it is insoluble in polar solvents, such as water.

If the third chain in the structure is replaced by a phosphoric group, which in turn can be combined with a variety of bases, the compound is a phospholipid. An example is *phophatidyl choline:*

$$\begin{array}{cccccccc}
H_2 & & H & & H_2 & & O & & & H_2 & & H_2 \\
| & & | & & | & & || & & & | & & | \\
C & - & C & - & C & - O - & P & - O - & [\ C & - & C & - N^+(CH_3)_3 \ ] \\
| & & | & & & & | & & & & \\
O & & O & & & & O^- & & & & \\
| & & | & & & & & & & \\
C=O & & C=O & & & & & & & \\
| & & | & & & & & & & \\
R_1 & & R_2 & & & & & & &
\end{array}$$

$$R_1 : (CH_2)_{16}-CH_3$$
$$R_2 : (CH_2)_{14}-CH_3$$

in which the base attached to the phosphate (shown in brackets) is choline.

There are a number of phosphatidyl cholines depending on the nonpolar side-groups. For instance, if the two nonpolar side-groups are palmitic acid the phospholipid is called *dipalmitoyl-phosphatidyl choline*. Sometimes the phosphatidyl cholines are called *lecithins*. They are the most abundant and most widely studied phospholipids. Another phospholipid is *phosphatidyl ethanolamine* with ethanolamine instead of choline as its base. A phospholipid that is not derived from glycerol is *sphingomyelin*,

$$\begin{array}{c}
\quad\quad\quad\quad H \quad\quad H \ \ H \quad\quad\quad\quad\quad\quad\quad O \quad\ \ H_2 \ H_2 \\
\quad\quad\quad\quad | \quad\quad | \ \ | \quad\quad\quad\quad\quad\quad\quad || \quad\ \ | \ \ | \\
CH_3-(CH_2)_{12}-C=C-C-C-CH_2-O-P-O-C-C-N^+(CH_3)_3 \\
\quad\quad\quad\quad\quad\quad\quad\quad | \ \ | \quad\quad\quad\quad\quad | \\
\quad\quad\quad\quad\quad\quad\quad\quad HO \ \ NH \quad\quad\quad O^- \\
\quad\quad\quad\quad\quad\quad\quad\quad\quad\quad | \\
\quad\quad\quad\quad\quad\quad\quad\quad\quad\quad O=C \\
\quad\quad\quad\quad\quad\quad\quad\quad\quad\quad | \\
\quad\quad\quad\quad\quad\quad\quad\quad\quad\quad (CH_2)_n \\
\quad\quad\quad\quad\quad\quad\quad\quad\quad\quad | \\
\quad\quad\quad\quad\quad\quad\quad\quad\quad\quad CH_3
\end{array}$$

An example of a steroid lipid is *cholesterol*:

Cholesterol and its derivatives seem to be essential components of at least some membrane structures (see section 2.4). Cholesterol most probably is also a precursor for many if not all steroid hormones.

Lipid bilayers
Lipids, and especially the phospholipids, are important structural factors in biological membranes. Although they cannot form polymers by covalent bonding, they can interact to form sheetlike structures. Pospholipids are *amphiphiles*, that is molecules which have, in addition to a relatively large nonpolar section, also a polar group. This type of molecules readily forms so-called bilayers, making them essential structural components of membranes. When in an aqueous environment, the long hydrophobic fatty acid chains form tails to the molecules which avoid contact with highly polarized water molecules. As a result, these parts tend to attract each other, excluding contact with the water; aggregates are formed which have these hydrophobic groups pointing to the inside, roughly parallel to each other and perpendicular to the plane of the aggregate and the polar (hydrophylic) groups pointing outward towards the water. Thus, a double molecular layer, callcd a *lipid bilayer* is formed (Fig. 2.10a).

It can be shown thermodynamically that this arrangement is the most favorable in terms of free energy; although the transfer of hydrophobic molecules to an aqueous environment goes with a decrease in enthalpy (energetically the molecules seem to prefer the water to an apolar environment), the large decrease in entropy causes the overall free energy change to be positive, and the hydrophobic molecules to prefer an apolar, nonaqueous environment. The molecules thus tend to aggregate spontaneously in the form of a bilayer. The driving force of this hydrophobic interaction or bonding thus is entropic.

In water the bilayer closes itself, forming vesicles with an interior separated from the outside (Fig. 2.10b). The closed vesicles form because if a free edge on a bilayer were exposed, some of the hydrophobic parts would be in contact with water; that would be thermodynamically unfavorable. Such closed vesicles form in the aqueous environment in and around cells. One can also prepare closed bilayer vesicles in the laboratory from various phospholipids. Lecithins from egg or soybean are very suitable for this purpose. Such laboratory prepared vesicles are called *lyposomes*.

Amphiphilic molecules with a single hydrophobic chain (as for instance dodecyl sulfate which has a long hydrocarbon chain) do not form bilayers

27

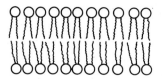

a

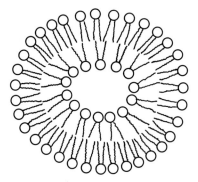

b

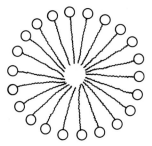

c

Fig. 2.10. Hydrophobic bonding of phospholipids (*a*) in a lipid bilayer; (*b*) in a lipid vesicle; (*c*) in a micelle.

but instead make *micelles*, globular aggregates with polar groups exposed to the surface and hydrophobic chains clumped together (Fig. 2.10c).

The behavior of the molecules in the lipid bilayer depends strongly on temperature. At sufficiently low temperature the molecules are rigid and the appearance is more like that of a hydrocarbon crystal. As the temperature is raised, sudden transitions occur at one or more temperatures; the hydrocarbon tails of the phospholipids begin to wiggle about and the bilayer behaves more like a two-dimensional liquid. The molecules are then capable of translational motion and can diffuse sideways (lateral diffusion). The rate of exchange of molecules between opposite monolayers remains very low, however. The transition temperatures depend on the lipid composition of the bilayer.

Bilayers can be composed of various phospholipids. The simplest arrangement would be a homogeneous structure in which there is a random distribution of the phospholipid molecules in both monolayers. However, because of the difference in curvature of the two bilayer surfaces and the presence of different solutions at both sides (inside and outside) of the bilayer the different phospholipids may segregate preferentially in the two monolayers. This asymmetric distribution of the phospholipids may be the result of a difference in thermodynamic stability between each type of lipid molecule and a given solution and curvature. Most naturally occurring membranes indeed are asymmetric. In the cytoplasmic membrane of the red blood cell, for example, the outer monolayer contains a mixture of phosphatidyl choline and sphingomyelin while, in contrast, the inner monolayer (facing the cytoplasm) has phosphatidyl ethanolamine and phosphatidyl serine.

2.4 Membranes

Membrane function

As we have mentioned before, membranes are an essential structural feature of life systems. They serve as selective barriers around cell and cell organelles; they form communication channels within the cell; many biophysical and biochemical processes require membranes.

Very broadly, one can define three types of membrane functions. First, as boundaries of cells and cell organelles, they create and maintain a definite chemical composition inside which can be quite different from the outside environment. They do this continuously by a combination of a selective passive diffusion and selective active (energy consuming) transport across the membrane. These processes will be discussed in more detail in Chapter 7; many details of the mechanisms by which the transport of material across the membranes occurs are not quite known, however. In organelles, such as the endoplasmic reticulum, membranes also serve as communication channels. Moreover, they are actively involved with processes like endocytosis and exocytosis.

Second, membranes can form a basis on which rapid chemical transformations, requiring an efficient supply of reactants and an efficient disposal of products and waste, take place. A variety of enzyme systems are associated with, or can be an integral part of membranes. These systems not only govern the transport of ions and/or molecules, but also govern the rate of various biological reactions. A substantial part of the energy-conserving reactions in biological systems requires membranes, thus demonstrating the essential character of membrane systems.

Finally, membranes occur as electrical insulation around fibrous extensions (called axons) of some nerve cells. Such an insulation is found in some vertebrate nerve cells and is called the *myelin sheath*. It originates from a satellite cell (called a Schwann cell) which winds itself, during the developmental stages, around the axon as indicated in Fig. 2.11. The entire cytoplasmic membrane

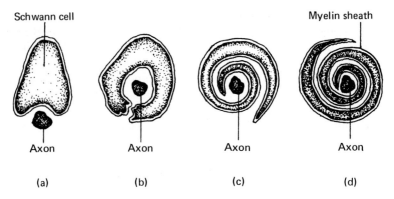

Fig. 2.11. Schematic representation of the myelin sheating of an axon, showing the progressive envelopment of the axon by the membranes of a Swann cell.

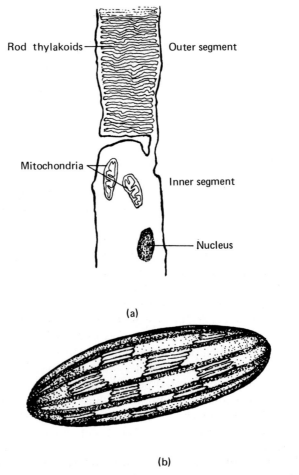

Fig. 2.12. Examples of the stacking of membrance portions in the form of flattened, disklike vesicles called thylakoids. (*a*) The rod receptor of a vertebrate retina; (*b*) a chloroplast.

Table 2.2. Composition of some cell membranes.

Membrane	Protein (%)	Lipid (%)	Carbohydrate (%)	Ratio of protein to lipid
Myelin	18	79	3	0.23
Plasma membranes				
blood platelets	33–42	58–51	7.5	0.7
mouse liver cells	46	54	2–4	0.85
human erythrocyte	49	43	8	1.1
amoeba	54	42	4	1.3
rat liver cells	58	42	(5–10)*	1.4
L cells	60	40	(5–10)*	1.5
HeLa cells	60	40	2.4	1.5
nuclear membrane of				
rat liver cells	59	35	2.9	1.6
retinal rods, bovine	51	49	4	1.0
mitochondrial outer				
membrane	52	48	(2–4)*	1.1
Sarcoplasmic reticulum	67	33	–	2.0
Chloroplast lamella,				
spinach	70	30	(6)*	2.3
Mitochondrial inner				
membrane	76	24	(1–2)*	3.2
Gram-positive bacteria	75	25	(10)	3.0
Halobacterium purple				
membrane	75	25	–	3.0
Mycoplasma	58	37	1.5	1.6

*Deduced from the analysis
Note: values in the table are percentages or ratios by weights.
Source: after G. Guidotti, Ann. Rev. Biochem. 41: 731 (1972). From "Biophysical Chemistry" by C.R. Cantor and P.R. Schimmel, copyright 1980; reprinted with permission of W.H. Freeman and Company.

of the Schwann cell thus forms an elaborate structure serving a specific function in connection with another cell, the neuron.

The myelin sheath is one demonstration of the flexibility of membranes, as their appearance is modified according to a specific function. Another example is the stacking of membrane portions in the form of flat disks as occurs in the rod cells of the retina (Fig. 2.12a) and in the grana of chloroplasts in a leaf cell (Fig. 2.12b). A common feature of these membrane stacks is that they contain light-receptor systems; the retina rod cell is an essential part of the process of vision (see Chapter 12) and the chloroplast has the necessary apparatus for photosynthesis (Chapter 6). These appearances reflect the structure of the membrane adapted to its specific function.

Membrane composition
Presently, there is little doubt that the phospholipid bilayer is a basic structure of all membranes. Biological membranes, however, are much more complex than simple bilayers; they contain components other than just phospholipids.

Table 2.2 lists the composition of some membranes taken from cells and cell organelles with different functions. In addition to the lipids there is generally a substantial amount of proteins present. Moreover, each membrane preparation contains in addition to lipid and protein some minor (10% or less) carbohydrate components (mostly sugars). Part of these carbohydrates are attached to lipids, forming the *glycolipids*. Glycolipids have hydrophobic tails like that of sphingomyelin but their hydrophilic end is composed of a variety of sugars joined together in linear or branching chains called oligosaccharides. They constitute only a minor fraction of the membrane and seem to be confined to the outer monolayer. Their biological function is not yet known. The other part is attached to proteins. The biological role of these *glycoproteins* is also not quite certain. They may have a function in the recognition process in receptor-mediated endocytosis.

A lipid other than phospholipid is cholesterol which is a major component in a number of membranes, such as in the red blood cell and in myelin. It embeds itself at the hydrophobic edge of the membrane, giving the membrane a more rigid structure (see Fig. 2.14).

As can be seen from Table 2.2, the protein is present in variable amounts. The amount is largely dependent on the function of the membrane. In myelin, for example, the ratio of protein to lipid is only about 0.2 whereas this ratio is more than 3 in mitochondrial membranes. Presumably, the only function of myelin is that of an insulator. Myelin, therefore, has no apparent need for a large amount of proteins. In mitochondria, on the other hand, the membrane is involved in enzymatic and transport reactions, implying an important role for proteins.

Membrane proteins
Membrane proteins fall into two broadly defined categories, depending upon how they are bound to the membrane. The first category comprises the *peripheral* proteins that are loosely bound to the membrane by electrostatic interactions. They can be removed from the membrane by relatively mild treatments and are stable in aqueous solutions with no tightly bound lipid material. Cytochrome *c*, an enzyme active in energy-converting electron transport in mitochondria (see Chapter 7) is an example. In the other category are the *integral* proteins which are embedded inside the membrane, often spanning it entirely. They are much more difficult to remove from the membrane, requiring treatment with organic solvents. These proteins are often isolated with bound lipid. Without the lipid they tend to aggregate and precipitate in an aqueous environment.

Figure 2.13 shows schematically the types and the arrangements of the membrane proteins. It seems reasonable to assume that the peripheral proteins interact with the membrane by contact with the integral proteins rather than with the lipids. The integral proteins occur in a wide variety of shapes. One shape is a rod-like spiral called an *α-helix* (see Chapter 4). Parts of the protein or the entire molecule can have this shape. Other shapes are more compact and bend together, forming globular structures. There is a wide variety of

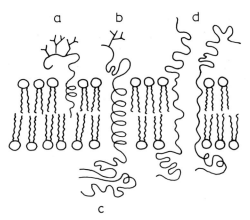

Fig. 2.13. Arrangement of membrane proteins; (*a*) integral protein partially in the membrane; (*b*) integral transmembrane protein; (*c*) peripheral protein; (*d*) channel forming proteins.

structures possible along these lines and, consequently, the proteins may have very different properties and behavior. These properties are reflected in the amino acid sequence of the membrane proteins. If a protein contains a region of predominantly hydrophobic amino acids, this region most probably resides inside the lipid bilayer. Regions with hydrophylic residues protrude outside the layer. Thus, proteins which span the entire bilayer must have a central hydrophobic region. Such proteins are called transmembrane proteins. Other integral proteins which are localized to one surface have a hydrophobic tail which anchors it in the membrane.

Because of the lateral fluidity of the lipid bilayer, proteins residing in the membranes can move in the plane of the membrane like particles in a solution or a suspension. Proteins can float individually or together as they are associated with other proteins. They can form more or less symmetric oligomers, helical structures or sheetlike structures. Presently, the quaternary structure of membrane proteins *in situ* is not well known but it is a problem which has a high priority in membrane research. Of particular interest is the possibility of channels through the membrane, caused by a single protein with a hole in it or by a composition of proteins that line a central cavity (see Fig. 2.13). These channels would allow transport of matter for which the bilayer by itself is impermeable.

Membrane structure
The picture that is emerging is a very dynamic one in which the membrane plays an *active* role, not only in transport processes but also in the cell biochemistry itself. A generally accepted working model is the so-called *fluid-mosaic* model proposed by S. Singer. This model is illustrated schematically in Fig. 2.14. The phospholipid bilayer forms a fluid matrix about 60 to 80 Å thick in which the various integral proteins are implanted. Some of these proteins protrude through both sides of the membrane, others through only one side. The proteins contain both hydrophobic and hydrophylic sections

33

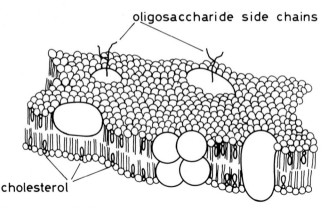

oligosaccharide side chains

cholesterol

Fig. 2.14. Fluid-mosaic model of a membrane.

and accommodate themselves thermodynamically in the corresponding parts of the lipid bilayer. Over the long range a protein can be randomly distributed but over the short range a specific distribution may exist for a particular protein. Lateral movement occurs but it is also possible that a protein may rotate about axes, either parallel or perpendicular to the plane of the bilayer. Such movements may be essential for some membrane functions.

2.5 Cell evolution

Prokaryotic and organellar similarities
Prokaryotic cells (bacteria) do not have the specialized organelles which we described in section 2.2 for the eukaryotic cells. Instead, they have more or less elaborate membrane systems which are linked to a variety of functions. These membrane structures, depending upon the kind of organism and metabolic condition, vary from more or less independent vesicle-like enclosures to stacks of lamellae which may or may not be continuations of the cytoplasmic membrane. The common feature of membrane structures in the prokaryotes, however, suggests relationships with the laminar structure inside the energy-converting mitochondria and chloroplasts. In fact, the inner membrane of mitochondria has many characteristics in common with the bacterial cell membrane, rather than with structures found inside the eukaryotic cell.

The presence of DNA and ribosomes inside mitochondria and chloroplasts, and the fact that mitochondria as well as chloroplasts divide (in some cases concurrent with cell division, in other cases more independently), gave impetus to the speculation that prokaryotic organisms and the energy-converting eukaryotic cell organelles may have a common origin. Circumstantial evidence tends to support such speculations; the organellar DNA, although incapable of programming for all proteins found in the organelle, suggests an independent existence at least early in evolution. The DNA in the organelle never assumes

34

the form of the chromosomes in the nucleus of the eukaryotic cell; it is much more similar to the bacterial DNA which is present in long closed strands.

Endosymbiosis

The demonstration of cellular endosymbiosis (the living of one cell inside the other) may be seen as a supporting indication. Endosymbioses occur with green algae inside certain protozoa; they are easy to detect because the characteristics of a eukaryotic green alga inside a (eukaryotic) host cell cannot be mistaken. Prokaryotic cyanobacteria (blue-green algae) may also become symbionts, although their presence as such inside a host cell is much more difficult to detect. Blue-green inclusions (cyanelles) are found in some amoeboid or flagellate eukaryotes but a definite proof that such cyanelles are in fact cyanobacteria cannot as yet be given.

Evolution of cells

Observations such as these suggest that endosymbionts may have been involved in the evolution of the eukaryotic cells. This evolution may have taken place according to the diagram shown in Fig. 2.15. According to this diagram, both prokaryotic and eukaryotic cells should have evolved from an ancestor cell. The archaebacteria discussed in section 2.1 may be the direct progeny of such ancestor cells. The present-day eukaryotic cells may have evolved because some of these cells developed a cytoplasmic membrane with the ability to take in particulate matter by some kind of endocytosis. Prokaryotes brought into the cell in this way may have developed into the organelles of the present-day eukaryotes, with the mitochondria and the chloroplast having retained some of their original independence. The present-day prokaryotes are, according to this suggestion, the progeny of those cells in which the cytoplasmic membrane did not develop this endocytotic ability. This suggestion was proposed in 1970 by R.Y. Stanier. There is no evidence to support or reject such as course of events. The fact that in present-day prokaryotic cells the cytoplasmic membrane lacks the plasticity to undergo the complete involutions necessary to take in objects with supramolecular dimensions may support the scheme of Fig. 2.15; processes like endocytosis have been observed only in eukaryotic cells.

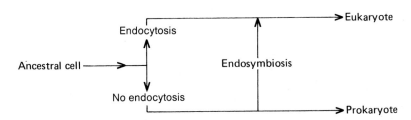

Fig. 2.15. Diagram showing the evolution of the cell: both eukaryotes and prokaryotes evolving from a common ancestral cell.

35

Bibliography

Allen, R.D. (1987). The Microtubule as an Intracellular Engine, *Sci. Am.* **256** (February), 26–33.

Bretscher, M.S. (1985). The Molecules of the Cell Membrane, *Sci. Am.* **253** (October), 86–90.

Cantor, C.R., and Schimmel, P.R. (1980). "Biophysical Chemistry Part I: The Conformation of Biological Macromolecules." W.H. Freeman and Cie, San Francisco.

Rothman, J.E. (1981). The Golgi Apparatus: Two Organelles in Tandem, *Science* **213**, 1212–1219.

Singer, S.J. (1981). Current Concepts of Molecular Organization in Cell Membranes, *Biochem. Soc. Trans.* **9**, 203–226.

Stanier, R.Y. (1970). Some Aspects of the Biology of Cells and their Possible Evolutionary Significance, in "Organization and Control in Prokaryotic and Eukaryotic Cells" (H.P. Charles and B.C.J.G. Knight, eds.) Cambridge Univ. Press, London.

Weber, K., and Osborn, M. (1985). The Molecules of the Cell Matrix, *Sci. Am.* **253** (October), 92–102.

Woese, C.R. (1981). Archaebacteria, *Sci. Am.* **244** (June), 94–106.

Woese, C.R., Debrunner-Vossbrinck, B.A., Oyaizu, H., Stackebrand, E., and Ludwig, W. (1985). Gram-positive Bacteria: Possible Photosynthetic Ancestry, *Science* **213**, 762–765.

3. Physical principles and methods in biology

3.1 The electronic structure of atoms

Basic principles

The sometimes spectacular advances made recently in biological research could not have occurred if it was not recognized that the behavior of life structures is based on physical and physicochemical principles. A good understanding of these principles is very important in biology. The interconnections between physics and biology indeed are the subject and the justification of biophysics. In this chapter we will discuss the physical and physicochemical principles as they apply to biology and also the physical techniques which have shown to be so useful in biological research. The great successes that came about by the application of these techniques reflect the fact that the composition and the behavior of biological structures are determined by these physical and physicochemical principles.

Molecules, including those in life systems, are made up of atoms. The electronic structure of molecules, therefore, cannot be understood without at least a basic knowledge of the electronic structure of atoms. It was not until the beginning of the twentieth century that some insight into the structure of atoms began to develop. This insight was gained in the first place from results of investigations of atomic emission spectra like, for instance, that of hydrogen. This spectrum consists of a number of series of lines whose wavelength separation and intensity decrease in a perfectly regular manner towards shorter wavelengths. Spectra from other atoms also show this regularity, though with an increasing degree of overlapping. It was Niels Bohr who first recognized the fundamental relation between this regularity and the structure of atoms. He started from a model proposed by Rutherford. In this model, an atom consists of a heavy nucleus with a positive charge Ze around which Z electrons rotate. Z is the atomic number of the atom in the periodic system of the elements and e is the absolute value of the charge of an electron. In order to explain the typical line spectra, Bohr had to postulate two basic assumptions:

1. Of the infinite number of orbits of the electrons about an atomic nucleus only a discrete number of orbits actually occur. These so-called quantum states are stationary; this means that in spite of the accelerated motion of the electrons and in contradiction to Maxwell's theory, no electromagnetic energy is emitted by the electrons while in these orbits.
2. Radiation is emitted (or absorbed) only by the transition of an electron

from one quantum state n_1 to another n_2. The energy difference between the two quantum states then appears as an emitted (or absorbed) light quantum, whose wavelength, λ, is given by

$$h\nu = \frac{hc}{\lambda} = E_{n_2} - E_{n_1} \qquad (3.1)$$

By making another postulate that in the quantum states the angular momentum I of the electron in its orbit is an integral multiple of Planck's constant divided by 2π,

$$|I| = M_e \nu r = n\,\frac{h}{2\pi}, \qquad n = 1, 2, 3, \ldots. \qquad (3.2)$$

(in which M_e and ν are the mass and the velocity of the electron and r is the radius of the orbit), Bohr was able to develop a theory which gave a surprisingly accurate explanation of the spectra of atoms and ions with a single electron. For atoms with more electrons, however, serious discrepancies showed up. Moreover, the quantum states themselves were hard to understand.

Atomic orbitals
Quantum mechanics solved these problems, at least in principle. The mathematical expression for Bohr's quantum states are the wavefunctions for the electrons in the atoms. These wavefunctions often are called *atomic orbitals*. They are the solutions of the Schrödinger wave equation $H\Psi = E\Psi$ (see appendix). For a hydrogen atom the solutions of the wave equation for the electron are determined by three parameters. One is the total energy given by

$$E = \frac{M_e e^4}{2n^2\hbar^2} \qquad (3.3)$$

in which e is the electrical charge of the electron, $\hbar$ is Planck's constant divided by 2π, and n is an integer which can have the value 1, 2, 3,... This integer n is called the *principal quantum number* and specifies the total energy of the electron. Another parameter is the angular momentum I associated with the orbital motion of the electron. Its magnitude is given by

$$|I| = \hbar \sqrt{l(l+1)} \qquad (3.4)$$

in which l is another quantum number, the *azimuthal* (or *orbital*) quantum number, which can assume values of 0, 1, 2,...,$n-1$. Thus, when the principal quantum number $n = 1$, the azimuthal quantum number can only have the value $l = 0$. Electrons for which $l = 0$ (and thus with zero angular momentum) are called s electrons. The wavefunctions for these electrons, as shown in Fig. 3.1, have spherical symmetry and are functions only of the radial coordinate. Electrons for which $l = 1, 2, 3, 4, 5,...$ are p, d, f, g, h,... electrons. The wavefunctions for these electrons also depend upon the angular coordinates (Figs. 3.1 and 3.2).

Then there is, as a third parameter, the orientation of the angular momentum which is also quantized. If we define a certain direction in space, for instance

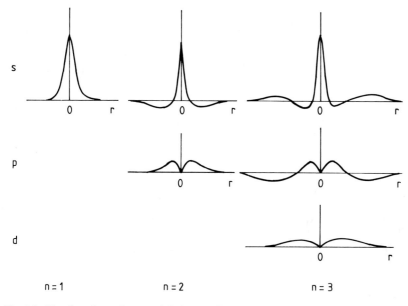

Fig. 3.1. Wavefunctions of s, p, and d electrons in hydrogen.

the z-axis of a Cartesian coordinate system (Fig. 3.2), it can be proved that the component of l in the z direction is given by

$$l_z = \hbar m_l \qquad (3.5)$$

in which $m_l = \pm 1, \pm 2, ..., \pm l$. This so-called *space quantization* clearly manifests

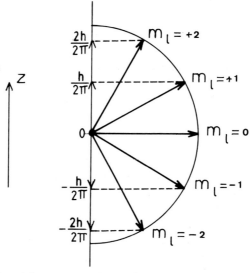

Fig. 3.2. Quantization of the axial component of the orbital angular momentum of an electron in an atom. The projection along the axis of an external field is determined by the magnetic quantum number m_l.

39

itself when the electron interacts with a magnetic field (see below). Therefore, m_l usually is called the *magnetic quantum number*. There are $2l + 1$ values of m_l. In the absence of a magnetic field, the energy levels of the states having different values of m_l are the same. These states are called *degenerate*; the number $s = 2l + 1$ is the *degeneracy* of the system.

Zeeman effect
Near the turn of the century the Dutch physicist Pieter Zeeman discovered that every line of the spectrum of atoms with one electron in a magnetic field is split into three equidistant lines. This phenomenon, since then called the Zeeman effect, is due to the fact that an electric charge that moves, like an electron in an orbit, has a magnetic moment. Using classical physics, it is easy to show that the orbital magnetic moment of the electron is proportional to its angular moment:

$$m_l = - \frac{e}{2M_e} l.$$
(3.6)

It turns out that a quantum mechanical derivation yields the same result. Therefore, using Eqs. 3.5 and 3.6, we can derive that the component of the magnetic moment in the z direction is

$$m_{lz} = - \frac{e}{2M_e} l_z = - \mu_B m_l$$
(3.7)

in which

$$\mu_B = \frac{e\hbar}{2M_e}$$
(3.8)

is the *Bohr magneton*.

The interaction energy of the electron's magnetic moment with a magnetic field is different for different values of the magnetic quantum number m_l. If l is 0, m_l can only be zero and there is no interaction with the magnetic field. This is evident since in that case there is no magnetic moment. If l differs from zero, the energy levels are split into as many levels as there are values of m_l (that is $2l + 1$) or, putting it differently, into as many levels as there are different orientations of the angular momentum. The degeneracy, hence, is lifted. However, transitions between those levels always span the same three energies. This is a result of a *selection rule* that stipulates that $\Delta m_l = \pm 1$ (see also below). The Zeeman splitting, therefore, occurs as a triplet.

Electron spin
In a first approximation the three parameters n, l, and m_l suffice to describe the spectral properties of atoms with one electron. It appeared, however, that a fourth quantity is necessary to describe certain aspects of some spectra, such as, for instance, the two closely spaced yellow lines in the spectrum of sodium. This quantity is related to an intrinsic angular momentum of the

electron itself and could be visualized as the angular momentum associated with the spinning of the electron about its own axis. Analogous to the orbital angular momentum, the magnitude of the spin angular momentum s is given by

$$| s | = h \sqrt{s(s+1)}. \tag{3.9}$$

The spin quantum number s, however, has the unique value of 1/2. Since an electron spin also can be visualized as a spinning electric charge, it has also a magnetic moment m_S. The relation between m_S and s, however, is different from that between m_l and l by a factor g, so that

$$m_s = - g \frac{e}{2M_e} s. \tag{3.10}$$

For a free electron the electronic g-factor has a value of 2.0023.

Just like the orbital angular momentum, the spin angular momentum has quantized components in the z direction.,

$$s_z = \hbar m_s, \tag{3.11}$$

specified by a spin magnetic quantum number m_S with possible values of +1/2 or –1/2. Thus the spin vector can have only two directions, roughly parallel and antiparallel.

The electron spin angular momentum can be added vectorially to the electron orbital angular momentum to give the total angular momentum

$$j = l + s \tag{3.12}$$

The total magnetic dipole moment, therefore is

$$m = m_l + m_s = \frac{e}{2M_e} (l + gs) \tag{3.13}$$

Multiplicity

In atoms with more than one electron, the angular momenta associated with the electron spin can be added vectorially (Fig. 3.3). The resulting total spin angular momentum is also quantized and determined by a total spin quantum number S. S is zero or an integer when we have an even number of electrons and is half-integer when we have an odd number of electrons. The orientation of this total spin angular momentum, again with respect to a reference direction, is quantized in the same way as the orientation of the orbital angular momentum (see Fig. 3.2); it is determined by a quantum number M_S which can have the values 0, ±1, ±2, ...,± S when S is an integer, or the values ±1/2, ±3/2, ±5/2,..., ±S when S is a half-integer. The total number of possible values of M_S, $2S+1$, is called the *multiplicity* of the system. Designations of the multiplicity are given in Table 3.1. Thus, in an atom with two electrons the total spin can be 0 when the electrons are paired (have antiparallel spins) and the multiplicity is 1. This is called a *singlet* state. The total spin also

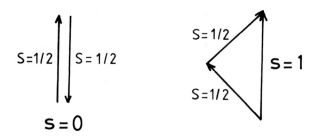

Fig. 3.3. Vector addition of antiparallel and "parallel" electron spin vectors.

can be 1 when the electrons are unpaired (have parallel spins; the electrons, in this case must occupy different orbitals to obey Pauli's exclusion principle which is described below). Then three orientations of the total spin vector in respect to an external magnetic field are possible and the state is a *triplet*. In atoms with an even number of electrons we can have singlet, triplet, etc. states; the states in atoms with an odd number of electrons can be a doublet, quartet, etc.

The total spin angular momentum S again can be added to the total orbital angular momentum L to give the quantized total angular momentum

$$J = S + L . \tag{3.14}$$

Stark effect
An electric field does not act on the magnetic moment associated with the total angular momentum J. The result of an electric field is a polarization of the atom. The resulting dipole moment is proportional to the electric field E:

$$\mu = \alpha E . \tag{3.15}$$

The proportionality constant α is called *polarizability*. It depends on the orientation of the "orbit", that is on the orientation of the total angular momentum J. The space quantization with respect to a magnetic field as described above also takes place in an electric field. This results in a shift of the energy which manifests itself in a shift of the spectral lines. The effect is known as the *Stark effect*. Its magnitude is proportional to both the dipole moment and the field strength. However, since the dipole moment is also

Table 3.1. Multiplicity of orientations of the spin vector.

S	M_S	Multiplicity
0	0	Singlet
1/2	+1/2, –1/2	Doublet
1	+1,0, –1	Triplet
3/2	+3/2, +1/2, –1/2, –3/2	Quartet

42

proportional to the electric field strength (equation 3.15), the spectral shift is proportional to the square of the electric field strength. This, together with the fact that the sign of the space quantization (the sign of M_j) has no effect on the energy level (the dipole moment is always induced in the direction of the field), makes the Stark effect qualitatively different from the more simple Zeeman effect.

The Pauli principle
The four quantum numbers n, l, m_l, and m_s determine the state of an electron in an atom or a molecule as described by its wavefunction. The *Pauli exclusion principle* dictates furthermore, that no two electrons in a system can be in the same detailed state. This means that no two electrons in a system can have the same set of quantum numbers. If two electrons are in the same orbital, having the same values for n, l, and m_l, they must differ in the spin quantum number m_s, or, speaking in a classical analogy, their spins must be antiparallel to each other.

In the ground state the electrons of an atom occupy the lowest energy levels allowed by the Pauli exclusion principle. In this way the electrons in the atomic orbitals distribute themselves over the quantum states as shown in Table 3.2. Quantum states differing from the ground state are called *excited states*.

Electronic transitions
The promotion (or degradation) of an electron from one quantum state to another is possible by the absorption (or emission) of a quantum of energy, satisfying relation 3.1. Such an event is called an *electronic transition*, and involves the change of one or more of the quantum numbers. There is a restriction, however, which follows from the quantum mechanical theory that dictates that the probability of the transition be extremely small unless the azimuthal quantum number l changes by +1 or by –1. This restriction is given by the selection rule $\Delta l = \pm 1$. Absorption and emission spectra of atoms such as hydrogen or the alkali vapors can be precisely explained by a detailed description along the lines pointed out above.

Table 3.2. Distribution of electrons in atomic orbitals.

Atomic shell	n	l	m_l	m_s	Designation	Number of electrons
K	1	0	0	–1/2,+1/2	1s	2
L	2	0	0	–1/2,+1/2	2s	2
		1	+1,0,–1	–1/2,+1/2	2p	6
M	3	0	0	–1/2,+1/2	3s	2
		1	+1,0,–1	–1/2,+1/2	3p	6
		2	+2,+1,0,–1,–2	–1/2,+1/2	3d	10

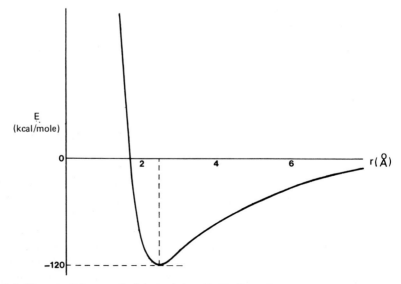

Fig. 3.4. The potential energy E of the ionic bond in NaCl as a function of the distance r between the ions Na^+ and Cl^-. The potential energy can be represented by $E = (q^2/r) + be^{-r/a}$ in which q is the ionic charge and a and b are constants.

3.2 The structure of molecules and molecular complexes

Interaction energy

The physics of molecules and molecular complexes describes many kinds of forces such as electron exchange forces, resonance forces, dipole forces, polarization forces, and Van der Waals forces. These names describe the conditions under which the forces are exerted but the kind of force is the same in all cases; it is the electrostatic force acting between the charged elementary particles in atoms which is described by Coulomb's law. Electromagnetic forces between moving charges are too weak to account for the formation of the structures under discussion.

A force exerted on a small object such as an atom or a molecule cannot be measured directly. We can, however, measure the energy by which, for instance, two atoms are held together by measuring the energy required to break the bond. If two atoms attract each other, their interactions represent a certain amount of potential energy. This potential energy reaches a minimum when the atoms have approached each other to such a distance that the attractive forces balance the repulsive forces. An example is the ionic bond between Na^+ and Cl^- in a molecule of NaCl vapor.* The minimum of the potential energy curve occurs when the attractive Coulomb force between

---

* In crystalline sodium chloride we cannot speak of an NaCl molecule since in the stable arrangement it is a three-dimensional crystal structure of Na^+ and Cl^- ions.

44

Table 3.3. Energy conversion table.

Wavelength, (nm)	200	400	800	1,600
Wavenumber,$=1/\lambda$ (cm^{-1})	50,000	25,000	12,500	6,250
Frequency (sec^{-1})	15×10^{14}	7.5×10^{14}	3.75×10^{14}	1.88×10^{14}
Joule	10^{-18}	5×10^{-19}	2.5×10^{-19}	1.25×10^{-19}
kcal/mol	144	72	36	18
eV	6.3	3.12	1.57	0.79

the two kinds of ions is balanced by the repulsive force between the nuclei (see Fig. 3.4).

The energy of the bond is the difference between the potential energy of the system with separated atoms and the potential energy of the system with bonded atoms. By convention the potential energy of a system of two atoms separated by an infinitely large distance is set at zero. Therefore, the bond energy is negative. Since the energy, rather than the forces, is usually the measurable quantity, it makes more sense to talk about interaction energy than about interaction forces. We can express the energy in Joule or electron volts per particle or, to conform with the chemists, in kilo-calories per mole (see Table 3.3).

Weak interactions
The criterion by which we make the distinction between strong and weak interactions is the extent to which thermal motion will disrupt the interaction. The average thermal energy is kT, in which k is Boltzmann's constant and T is the absolute temperature (in degrees Kelvin). At body temperature (310 °K) this is of the order of 2.5×10^{-2} eV/particle (about 0.6 kcal/mole). Strong interactions have a value many times greater than this and are, thus, unlikely to be disrupted by thermal motion. The primary structure of biological macromolecules is determined by such strong interactions. The higher-order structures are determined by weaker forces. Weak interactions are of the same order of magnitude as kT. These will be disrupted first when the molecule is heated, resulting in the loss of quaternary, tertiary, and secondary structure in that order.

Weak interactions between atoms and molecules cause a failure of the gas law of Boyle and Gay-Lussac, leading to the Van der Waals equation,

$$[P + (a/V^2)] (V - b) = RT \qquad (3.16)$$

in which a and b are (to a first approximation) constants. Weak interactions, therefore, are often called Van der Waals interactions or Van der Waals forces. These forces, electrostatic in nature, can be described as interactions in which electric dipoles are involved.

Dipoles
A bond between two atoms with different electronegativities (different affinities for negative charges or electrons) always results in a dipole moment. A typical example is the OH bond. The electronegativity of the oxygen atom draws

45

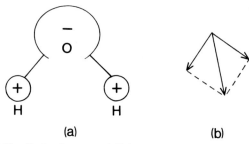

(a) (b)

Fig. 3.5. (a) The dipole of water and (b) its vector representation.

the electron from the hydrogen atom toward the oxygen and the result is a system in which equal charges of different sign are separated. The dipole moment of the OH bond is 1.60 D. [1 D (= 1 debye) is equal to 10^{-18} esu cm].

Water consists of two such bonds in which one oxygen and two hydrogen atoms participate. The resultant dipole moment of water is 1.85 D. This means that the two OH bonds must make an angle of 105° with each other. Only then can vector addition result in a total dipole moment as determined (Fig. 3.5).

Dipole moments for some molecules are given in Table 3.4. These moments give information about (1) the extent to which a bond is permanently polarized and (2) the geometry of the atoms, especially the angle between them (as we have seen with water). The fact that the dipole moment of carbon dioxide is zero in spite of the difference in electronegativity between carbon and oxygen indicates that the molecule is linear,

$$O = C = O$$

with the two dipole moments of the C=O bonds canceling each other out. Other examples are given in Fig. 3.6: Benzene (a), p-dichlorobenzene (b), and 1,3,5 trichlorobenzene (c) must be planar to account for their zero dipole moment. The two hydroxyl groups in p-dihydroxybenzene (d), however, must make an angle with the plane of the ring in order to account for the 1.64 D value of its dipole moment.

Dipole interaction
Dipoles interact with an electric field by Coulombic forces, which tend to

Table 3.4. Dipole moments.

Molecule	Moment (Debye)
HCl	1.03
H_2O	1.85
1,3,5-trichlorobenzene	0
p-dihydroxybenzene	1.64
Glycine	15.1
Egg albumin	252

46

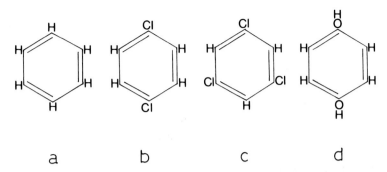

Fig. 3.6. (*a*) benzine; (*b*) *p*-dichlorobenzene; (*c*) 1,3,5-trichlorobenzene; (*d*) *p*-dihydroxybenzene.

align the dipole in the direction of the field. If the field originates at a point charge q, the force acting on the aligned dipole can be calculated by application of Coulomb's law, under the assumption that the distance between the point charge and the dipole is large as compared with the distance separating the two charges of the dipole. From such a calculation it follows that the energy of this type of ion-dipole bond is proportional to the inverse square of the distance (whereas the mutual energy of two ions in an ionic bond is inversely proportional to the first power of the interionic distance). Using a distance of 2.73 Å between a K^+ ion and a water molecule in the first hydration sphere and a dipole moment of 1.85 D for the water dipole, one can calculate that the binding energy of the first hydration sphere of K^+ is about 0.7 ev (17 kcal/mole). This is about 15% of the (monovalent) anion-cation bond at the same distance. Thus, hydration of a monovalent ion such as K^+ is not much disturbed by thermal motion; at least not in the first hydration sphere. However, since the binding energy decreases with the square of the distance, the orienting force of an ion on water dipoles must become quite inefficient beyond a layer of three or four molecules. Dipole-dipole interactions contribute to the diminution of the binding energy beyond the first few hydration spheres. Dipole-dipole interaction diminishes with the third power of the dipole distance.

Permanent dipoles are characteristic of molecules in which relatively electropositive atoms are bound to relatively electronegative ones (that is when the dipoles of several bonds do not cancel each other out such as with the examples given above). All molecules acquire dipoles by relative displacement of their positive and negative charges when placed in an electric field. The magnitude of such induced dipoles depends on the polarizability α of the molecule. The interaction of induced dipoles is a second-order effect; when the dipole is induced by a field from a point charge the interaction energy is, as can be calculated easily, proportional to the fourth power of the inverse distance. Dipole-induced dipole interaction energy is proportional to the sixth power of the inverse distance.

The hydrogen bond
Dipole-dipole interactions (between polar molecular groups) and dipole-

47

induced dipole interactions are important for the secondary and higher-order structure of proteins. Another type of electrostatic interaction plays a very important role in biology; this is the already mentioned hydrogen bond. Hydrogen, since it has only one 1s electron, can form only one covalent bond. In some situations, however, hydrogen can bond with two other atoms, instead of just one. The additional bond, which is much weaker than a normal covalent bond, is due to electrostatic attraction between the proton and some electronegative element of small atomic volume, such as N, O, or F. The proton, which is slightly isolated from the valence electron involved in the covalent bond, is extremely small; its electrostatic field, therefore, is intense and a bonding can occur due to the attraction of the positive proton for the electrons of the bonded atom.

The distances involved in this case (2–3 Å) are in the same order of magnitude as the dipole distances (about 1 Å or slightly smaller). The dipole-dipole approximation (in which the distance between the poles is assumed large compared to the dipole distance), therefore, cannot be applied; more detailed calculation shows that the interaction energy is some 5 to 10 times greater than the "close-range" dipole-dipole interaction. The bonds are indicated by dashed lines in order to distinguish them from the covalent bond in which the hydrogen is engaged. The O–H---O hydrogen bond is about 0.2 eV (5 kcal/mole), the O–H---N is somewhat weaker.

Hydrogen bonds are responsible for the structure of water and ice (Fig. 3.7). Te dimer of formic acid

$$
\begin{array}{ccc}
& \text{O–H----O} & \\
\text{H–C} & & \text{C–H} \\
& \text{O----H–O} &
\end{array}
$$

is a result of hydrogen bonds. In biology hydrogen bonds are, to a major

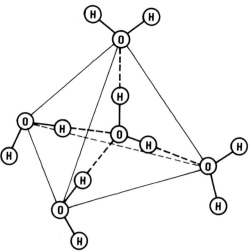

Fig. 3.7. The structure of water due to hydrogen bonding.

degree, responsible for the tertiary structure of proteins and nucleic acids. We have met them already in the structure of DNA (Chapter 2) and we will see (Chapter 4) how they are instrumental in the processes of replication, transcription, and translation.

Ionic and covalent bonds

Molecules are held together by strong interactions, or bonds, between the atoms. The bond energies are well in excess over thermal energies, as can be seen from Table 3.5. For polar molecules, such as in NaCl vapor, the bond can be described as the electrostatic attraction between a positive and a negative ion. The explanation of nonpolar molecules, such as CH_4, is somewhat more complicated (see below). For both types of molecules the valence (the number of atoms that each atom can bond) is usually the same. Oxygen, for example, binds two potassium atoms in the polar compound K_2O and two ethyl groups in the nonpolar compound $(C_2H_5)_2O$.

Since molecules are made of atoms they can be considered as systems of charged particles which assume a certain equilibrium configuration resulting from mutual attraction and repulsion of these particles. Therefore, the only thing we have to do in order to calculate the properties of the system is to set up an appropriate Hamiltonian which includes all these interactions and solve the Schrödinger equation. This yields, in addition to the wavefunctions ψ, the possible values of the energy E of the system and a number of physical quantities that can be calculated from ψ through use of the appropriate equations. The differential equation which we have to solve, however, is an extremely complex one, including all the electrons. Moreover, the terms describing the mutual interactions of the electrons bring about unsolvable mathematical problems. In practice, only systems with one electron can be rigorously solved.

Table 3.5. Dissociation energies of typical bonds.

Bond	Dissociation energy (eV)
H–H	4.40
C–C	2.55
C–H	3.80
C–N	2.13
C–N (peptide)	3.03
C=C	4.35
C=O	6.30
— C≡C—	5.35

Molecular orbitals

Fortunately, a number of approximation procedures have been developed which result in approximate solutions of the equation. Such solutions are useful in as much as they enable us to interpret observed experimental facts in terms of the appropriate fundamental physical quantities. The *molecular orbital* method is such an approximation procedure, suitable for molecular systems. It is essentially a method in which molecular orbitals, wavefunctions for single electrons in a molecule, are constructed by the fusion of atomic orbitals, the wavefunctions for single electrons in the atoms which form the molecule. If such a fusion is carried out by linear combination of the atomic wavefunctions, the procedure is called the MO–LCAO (molecular orbital-linear combination of atomic orbitals) approximation. The MO-LCAO procedure can be illustrated with the molecular orbital of the hydrogen molecule.

In Fig. 3.8, the interactions of two hydrogen atoms A and B are shown. If the atoms are far apart the electron of each of the two atoms occupies an atomic orbital which is the single 1s orbital, $\Phi_A(1s)$ and $\Phi_B(1s)$. When the atoms are brought together the atomic orbitals coalesce, and if we adopt the principle that the molecular orbital can be constructed from a linear combination of atomic orbitals (LCAO) we obtain

$$\psi_{(MO)} = C_1\Phi_A(1s) + C_2\Phi_B(1s) \tag{3.17}$$

Since the molecule is completely symmetrical C_1 must be equal to $\pm\, C_2$. We thus obtain two possible MOs,

$$\psi_g = \Phi_A(1s) + \Phi_B(1s), \quad \psi_u = \Phi_A(1s) - \Phi_B(1s) \tag{3.18}$$

In Fig. 3.9 cross sections of these orbitals are sketched. According to the Pauli principle, each orbital can have two electrons provided that they have antiparallel spins.

The MO labeled ψ_g does not change sign upon inversion through the center of symmetry in the midpoint between the two nuclei. This wavefunction,

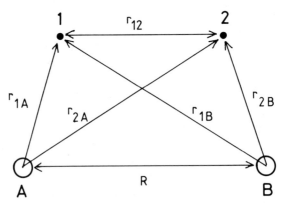

Fig. 3.8. The mutual interactions of two hydrogen atoms A and B. *R* is the internuclear distance, r_{ij} are the distances between nuclei and electrons.

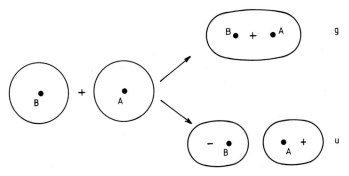

Fig. 3.9. Two molecular orbitals g and u of the hydrogen molecule formed by two atomic orbitals of hydrogen.

therefore, is called *even* (the subscript g stands for *gerade* which is German for even). The other wavefunction ψ_u does change sign upon inversion and is, therefore, *odd* (u stands for *ungerade* which means odd). The even molecular orbital ψ_g leads to a buildup of electronic density between the two nuclei, the odd molecular orbital has a node between the nuclei. In other words, the even orbital can be, and in the ground state is, occupied by electrons which are shared by the nuclei while in the odd orbital the absence of electron density between the nuclei makes them repulse each other. Therefore, in this case the even orbitals are bonding and the odd orbitals are antibonding. Each of the orbitals can be occupied, in accordance with the Pauli principle, by two electrons. Thus, in the covalent bond of the hydrogen molecule, the bonding orbital is occupied by two electrons with antiparallel spins. Excitation of the molecule leads to the promotion of one electron to the antibonding orbital.

An elaboration of this leads to a quantum mechanical (MO–LCAO) approximation for the description of the covalent bond. It starts from the system of nuclei and adds the electrons one by one, taking into account the Pauli exclusion principle. The procedure is very similar to the procedure by which atoms can be built up by adding the s, p, d, etc., electrons in the lowest atomic orbitals, as dictated by the Pauli principle.

Orbital quantum numbers
As with atomic orbitals, the molecular orbitals are determined by quantum numbers. For diatomic molecules the molecular orbital quantum numbers are the principal quantum number n (with $n - 1$ being the number of nodes in the wavefunction) and a quantum number λ which determines the component of the angular momentum along the internuclear axis. The quantum number λ thus has an analogy with the magnetic quantum number m_l in atoms. Orbitals for which $\lambda = 0, 1, 2,...$ are called σ, π, δ,... orbitals and electrons occupying such orbitals are called σ, π, δ,... electrons. In diatomic molecules the wavefunctions for σ orbitals are symmetric with respect to the axis connecting the two nuclei. π orbitals, since they have an angular momentum of unity, are antisymmetric about this axis. The higher orbitals have more complicated symmetry patterns.

51

In addition to σ, π, δ,... orbitals in molecules, there are so-called n orbitals that resemble atomic orbitals embedded in the molecule. An n electron is confined to the neighborhood of one nucleus and interacts only weakly with the other nuclei. The wavefunctions of the n orbitals in a molecular atom are not very different from their counterparts in the isolated atom.

Orbitals generated by the fusion of s orbitals normally are σ orbitals, such as the one in the hydrogen molecule. p orbitals can form σ as well as π orbitals; an example is the nitrogen molecule. Nitrogen has three 2p electrons; their orbitals have a nodal point at the nucleus and have a butterfly-like configuration, as sketched in Fig. 3.10. The three orbitals are at right angles to each other and can be labeled p_x, p_y, and p_z. If two N atoms come together the p_x orbitals fuse together to form a σ orbital, as sketched in Fig. 3.10a. These are the most stable orbitals that can be formed from the p orbitals. The molecular orbitals formed from the p_y and p_z atomic orbitals of the two N atoms have a distinctly different form. The sides of the p orbitals coalesce to form "streamers" of charge density, one above and one below the internuclear axis (Fig. 3.10b). The angular momentum now has a component along the internuclear axis, the quantum number $\lambda = 1$, and the orbitals are π orbitals. In this case the odd orbital (normally occupied) is bonding and the even orbital is antibonding.

Hybrid orbitals

In multiatomic molecules the atomic orbitals often combine to form so-called *hybrid orbitals* before they form molecular orbitals. A classical and biologically important example is carbon. Carbon is bivalent in the ground state; it has two unpaired electrons. In compounds, however, carbon has four valences often directed to the corners of a regular tetrahedron. The hybrid orbitals explain this phenomenon. In the ground state carbon has two unpaired electrons in atomic orbitals p_x and p_y. In order to acquire the four valencies, one of the 2s electrons is promoted to a 2p state. The atom then obtains four

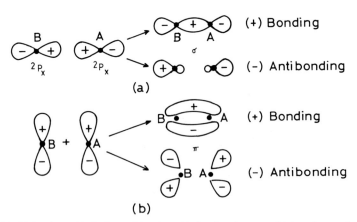

Fig. 3.10. Molecular orbitals formed by the orbitals of p electrons: (*a*) bonding and antibonding σ orbitals; (*b*) bonding and antibonding π orbitals.

unpaired electrons, a 2s electron, and three 2p electrons. This promotion requires approximately 2.8 eV, but this energy is more than compensated for by the binding energy of the four bonds. The four orbitals are now combined with each other in hybrid orbitals. An extremely stable way to do that is to combine the 2s orbital with each of the p orbitals in the following manner:

$$\psi_1 = 1/2(\Phi_s + \Phi_{px} + \Phi_{py} + \Phi_{pz})$$

$$\psi_2 = 1/2(\Phi_s + \Phi_{px} - \Phi_{py} - \Phi_{pz})$$

$$\psi_3 = 1/2(\Phi_s - \Phi_{px} + \Phi_{py} - \Phi_{pz})$$ (3.19)

$$\psi_4 = 1/2(\Phi_s - \Phi_{px} - \Phi_{py} + \Phi_{pz})$$

The resulting four hybrid orbitals designated as sp³ (each having a shape shown in Fig. 3.11a), are directed to the corners of a regular tetrahedron as shown in Fig. 3.11b. In this configuration the negative charges in the orbitals avoid each other in the maximum as much as possible. If the four orbitals are now combined with the 1s orbitals of four hydrogen atoms, four σ molecular

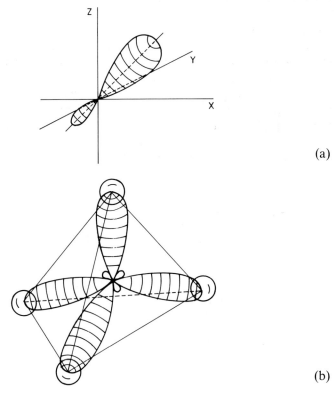

(a)

(b)

Fig. 3.11. (*a*) A sp³ hybrid ortibal: (*b*) the molecular orbitals of methane formed from sp³ hybrid orbitals.

orbitals are formed in the methane molecule CH_4.

Conjugated double bonds (resonance)
Hybridization of atomic orbitals in carbon can also lead to other configurations. One example is the formation of three sp^2 hybrids (trigonal hybrids):

$$\psi_1 = (1/\sqrt{3})\Phi_s + (\sqrt{2}/\sqrt{3})\Phi_{px}$$

$$\psi_2 = (1/\sqrt{3})\Phi_s - (1/\sqrt{6})\Phi_{px} + (1/\sqrt{2})\Phi_{py} \qquad (3.20)$$

$$\psi_3 = (1/\sqrt{3})\Phi_s - (1/\sqrt{6})\Phi_{px} - (1/\sqrt{2})\Phi_{py} .$$

Here the 2s, $2p_x$, and $2p_y$ orbitals are mixed, leading to three coplanar orbitals separated over an angle of 120° and leaving the fourth AO, the p_z, perpendicular to the plane of the others. This form of hybridization leads to the formation of nonlocalized molecular orbitals, an example being benzene. The three sp^2 hybrids are fused with each other and with the 1s orbital of the H atoms to form localized σ orbitals which are in the plane of the molecule (Fig. 3.12a). The p_z orbitals, which extend above and below the molecular plane (Fig. 3.12b), then fuse together to form π orbitals above and below the plane (Fig. 3.12c). Of the six possible orbitals three are bonding and three are antibonding. In the ground state only the three bonding (lowest energy) orbitals are occupied, each with two electrons which, in accordance with the Pauli exclusion principle, must have antiparallel spins. The six electrons in these orbitals are no longer localized to a particular atom. They can move freely in the doughnut-like spaces.

Because each electron has more space to occupy as a result of the delocalized orbitals their energy levels are lower (this is a consequence of the uncertainty principle). This extra energy (for benzene 1.8 eV or about 41 kcal/mole) contributes to the strength of the bond and is called *resonance energy*. If the structure of a molecule is seen as being made up of the superposition of various valence-bond structures, we can say that the actual structure, for instance of benzene, is formed by the "resonance" of two Kekulé structures

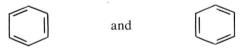
and

This type of bond structure is called a *conjugated double-bond* structure. Many biologically important molecules have such conjugated structures which give the molecule certain easily recognizable spectroscopic characteristics (such as the chromophores of cytochrome, hemoglobin and myoglobin, and chlorophyll).

Molecular electronic transitions
If two 2p atomic orbitals coalesce into π molecular orbitals, two of these are possible; a bonding orbital (π) and an antibonding orbital (π^*). The energy

54

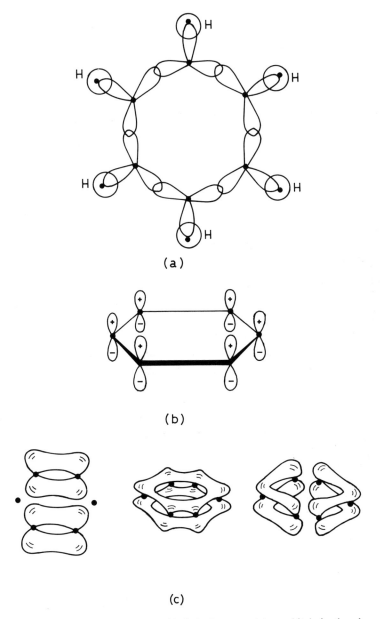

(a)

(b)

(c)

Fig. 3.12. The formation of molecular orbitals in benzene; (*a*) σ orbitals in the plane of the molecule; (*b*) the p_z atomic orbitals; (*c*) the three bonding π orbitals formed by the p_z orbitals.

levels, with respect to the energy level of the atomic orbital, are as given in Fig. 3.13. In the ground state the two electrons usually occupy the lower π orbital, having antiparallel spins according to the Pauli principle. Upon absorption of an energy quantum (light) of the right size, a transition may occur which promotes one electron from the π orbital to the π* orbital. The

Fig. 3.13. The energy levels of the π orbitals formed by two p orbitals. In the ground state the two electrons having opposite spins occupy the lower energy orbital.

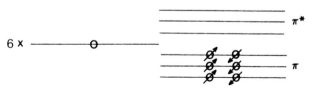

Fig. 3.14. The energy levels of the π orbitals formed by six p orbitals, as they occur in benzene. In the ground state the three lower energy levels are occupied each by two electrons having opposite spins.

energy involved in such a transition, in the visible or near uv spectral region, is far less than the energy involved in atomic electronic transitions, in the far uv; this is due to splitting of the p orbital energy level. This proliferation of energy levels is even more pronounced when a system similar to that of benzene is considered (Fig. 3.14). In this system any one of the six electrons in the ground state occupying three π orbitals can be promoted to three π^* orbitals. It is then clear that the more extended the conjugated system the broader the absorption bands and the more they will extend toward the red

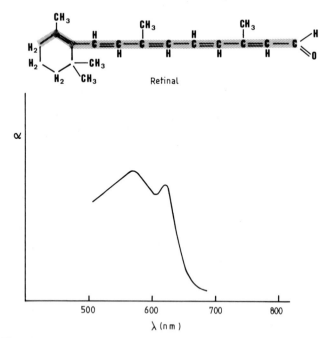

Fig. 3.15a. The absorption spectrum of retinal. The chemical structure of the molecule is also shown. The hatched areas designate the π systems in the molecule.

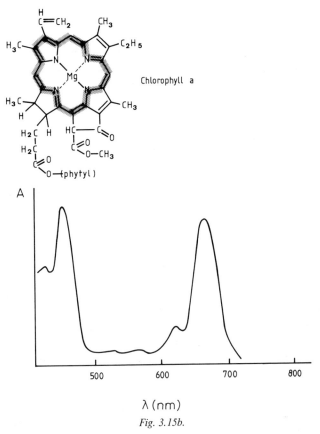

Chlorophyll a

A

500	600	700	800

λ (nm)

Fig. 3.15b.

Fig. 3.15b. The absorption spectrum of chlorophyll *a*. The chemical structure of the molecule is also shown. The hatched areas designate the π systems in the molecule.

part of the spectrum. This is illustrated in Fig. 3.15 a and b, which compare the spectrum of the eye pigment retinal with that of chlorophyll *a*.

Transition diagrams

Transitions from one molecular orbital to another, e.g. from π to π^* or from n to π^*, are called $\pi\pi^*$ or $n\pi^*$ transitions. In Fig. 3.16a a sequence of transitions is illustrated with an energy level diagram. In such a diagram the levels given are the energy levels of the electrons in their respective orbitals. The sequence shows (1) the promotion of a π electron to a π^* orbital, (2) the transfer of an electron from the n orbital to the partly vacated π orbital (the net result of these two steps is the promotion of an n electron to a π^* orbital), and (3) the fall of the π^* electron into the vacancy in the n orbital. The same events are more conveniently and correctly represented by the so-called transition diagram (Fig. 3.16b). In this diagram the energies of the ground state and the various excited states of the molecule as a whole are depicted. Here state means the state in which an electron (no matter which electron)

57

has been promoted from the n orbital to the π^* orbital. Thus, the sequence is (1) a $\pi\pi^*$ excitation, (2) a $\pi\pi^*$-*to-nπ^** interconversion, and (3) a deexcitation from $n\pi^*$ to ground state.

$\pi\pi^*$ and $n\pi^*$ transitions

The most intense transitions are those in which a strong dipole oscillation is involved and where the orbitals have a good spatial overlap. This is the case for $\pi\pi^*$ transitions which are, therefore, highly favored and cause intense absorption bands. This can be understood if one considers that in the $\pi\pi^*$ excitation the excited π^* electron keeps a strong coupling with its partner in the π orbital, and that the π and the π^* orbital, although differing in symmetry, occupy the same region in the molecule.

In an $n\pi^*$ transition, an electron moves from a localized orbital to a delocalized orbital. The spatial overlap of the two orbitals, therefore, is very poor and the transition probability is correspondingly low. As a result the absorption bands of $n\pi^*$ transitions are about a hundred-fold less intense than those of the $\pi\pi^*$ transitions. The intrinsic lifetimes of the $n\pi^*$ states are greater than those of the $\pi\pi^*$ states by the same factor. However, $n\pi^*$ states have a high degree of polarization due to the large electron displacement attending the transition. This may be the reason that once an $n\pi^*$ state has been attained the molecule becomes very reactive as an electron donor or an electron acceptor. Oxidation-reduction reactions are very common in biology and $n\pi^*$ states may very well be involved.

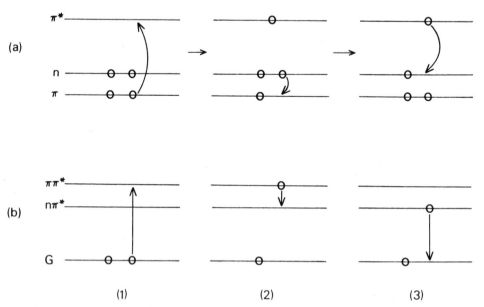

Fig. 3.16. (a) The energy level diagrams and (b) the transition diagrams of the transitions $\pi\pi^*$ (1), $\pi\pi^*$- to- $n\pi^*$ interconversion (2), and $n\pi^*$- to-ground transition (3).

The triplet state

When the spin of the excited electron is reversed in a transition from a singlet state a triplet state is generated. Such a spin reversal in the transition from the ground state is very improbable since the antiparallel electrons are strongly coupled in the ground state. From a first excited state a spin reversal can occur more easily since the spin-spin coupling has become looser. The principal magnetic forces that will disrupt spin-spin coupling are the orbital magnetic moments of electrons and atomic nuclei. Thus, the greater the spin-orbital coupling and the weaker the spin-spin coupling, the more likely a singlet-triplet transition can occur.

A greater spin-orbital coupling and a weaker spin-spin coupling are especially marked in $n\pi^*$ states, as contrasted with $\pi\pi^*$ states. Further, the $n\pi^*$ states are intrinsically longer lived than the $\pi\pi^*$ states. Entry into a triplet state will, therefore, more likely occur from the former than the latter state. Triplet states have lower energy than singlet states, because the excited electron and its ground state partner have parallel spins. The more closely coupled the two electrons, the larger the energy difference between the singlet and the triplet state. Thus, the energy gap between a $\pi\pi^*$ and a $\pi\pi^T$ state is considerably larger than that between an $n\pi^*$ and an $n\pi^T$ state (see Fig. 3.17). The lifetime of a triplet state is four to five orders of magnitude longer than that of a singlet state. Triplet states, therefore, are often implicit in photochemistry. There is, for instance, good evidence that triplet states are involved in the *in vitro* photochemistry of chlorophyll. Triplet states seem to play no role in the function of chlorophyll *in vivo*, however (see Chapter 6).

3.3 Absorption and emission spectroscopy

Electronic transitions in molecules occur with absorption or emission of light (from the uv to the near infrared spectral region). The emission of light is called *fluorescence* when it accompanies "allowed" radiative transitions such as the return from a $\pi\pi^*$ state to the ground state and is short lived (in the order of 10^{-8} sec for molecules like chlorophyll). Light emitted from much longer lived, so-called metastable states (such as triplet states) is called

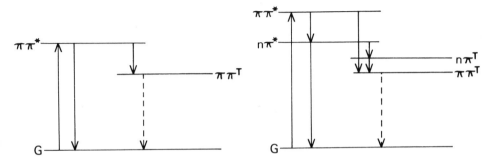

Fig. 3.17. Transition diagrams involving $\pi\pi^*$ to $\pi\pi^T$ and $n\pi^*$ to $n\pi^T$ transitions.

phosphorescence. When light is emitted from the first excited singlet state after a back transition from a metastable state it is called *delayed fluorescence* or *luminescence*.

Line and band spectra

When the energy levels of the quantum states are well-defined and separated, as is the case in atoms, the absorption and emission spectra show sharp and intense lines or narrow bands. In molecules, however, there is a greater proliferation of quantum states of different energy. The electrons of atoms brought together can interact with each other and with more than one nucleus, thus becoming components of a larger system. As a result the original energy levels are split up into numerous sublevels. The relative movements, vibrational and rotational, of the nuclei with respect to each other also contribute to this proliferation of quantum states. Although one can still recognize the major electronic configuration in the ground state and the excited states (electronic states), each of these is subdivided in a manifold of substates reflecting the finer details of the interactions inside the molecule. The possibilities of transitions of different energy are thus substantially greater and the line spectra become band spectra.

Vibrational and rotational states

The relative motions of the nuclei, vibrational and rotational, result in substates which are, although of course much closer together than the electronic states, also quantized. An electronic state thus is subdivided in a set of quantized vibrational substates and each of these, in turn, is subdivided in a set of rotational "subsubstates". The role of nuclear vibrations in molecular spectroscopy can best be explained by discussing, as an example, a model of an idealized diatomic molecule such as the one shown in Fig. 3.18a. The covalent bond determined by the orbitals of the binding electrons holds the system together while the repulsive electrostatic force between the nuclei tends to pull it apart; hence, the system vibrates. Let us assume that we can consider the system as a harmonic oscillator (which it would be when the nuclei were particles connected by a spring obeying Hooke's law). Then the potential energy of the vibration is a parabolic function, given by the bold line in Fig. 3.18b, of the distance r separating the nuclei. Classically, such a system could exist anywhere within the domain bordered by the parabola but not outside this domain. Vibration changes the internuclear distance along line segments in the parabola which, because of the conservation of energy, are parallel to the r axis. All line segments between the two arms of the parabola are possible, or in other words, the total energy of the system can assume a continuum of values. On the molecular level, however, we have to use the total energy of the system to construct a Hamiltonian operator which, acting on the wavefunction ψ, determines the Schrödinger equation. The only solutions (wavefunctions) of this equation, and hence the only possible states of the system, are those in which the energy E_{vib} is given by

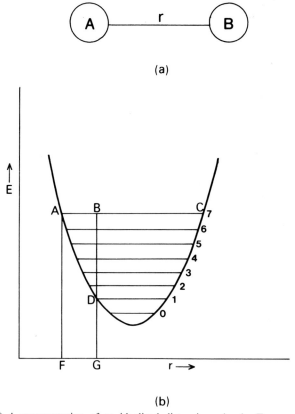

(a)

(b)

Fig. 3.18. (a) A representation of an idealized diatomic molecule. Two masses A and B are held together by a weightless spring. *(b)* The potential energy curve of the system shown in *(a)*. When the system vibrates the energy can have values within the area bounded by the curve (for instance between A and C). In a quantum mechanical system only discrete levels 0,1,2,... are possible. At points on the curve (for instance point A) the system is at rest and the energy is all potential energy (AF). When the system moves (for instance in point B) the energy comprises kinetic energy (BD) and potential energy (DG).

$$E_{vib} = h\nu_0(v + 1/2) \qquad (3.21)$$

in which h is Planck's constant, ν_0 the characteristic frequency of the oscillator, and v the vibrational quantum number (which can have the integer values 0,1,2,...). In Fig. 3.18b the energy levels for a number of values of v are drawn as thin lines. In classical terms we could describe the state of the molecule in the seventh vibrational level ($v = 7$) by the line segment AC. At point A or C, the velocity of vibration would be zero and the vibrational energy would only be potential. In B, however, the system moves and BD represents the kinetic part of the energy. Any point on the diagram of the horizontal line segments would then represent an instant at which the nuclei have a certain position and momentum. This statement needs to be modified, of course, owing to Heisenberg's uncertainty principle.

61

Franck-Condon principle

Potential curves can be drawn for the ground states and the excited states. An electronic transition always starts from a vibrational level in one state and terminates in a vibrational level in another state. Such transitions are shown in Fig. 3.19 for a diatomic molecule. Transitions from the ground state to an excited state follow from absorption of the appropriate quanta. Since the time in which a transition occurs is much shorter than the time of a nuclear vibration, during a transition neither the momentum of the nuclei nor their relative positions will change. This is a verbal statement of the *Franck-Condon principle*. Graphically, this principle can be expressed by the statement that, in plots like the one given in Fig. 3.19 only vertical arrows represent possible transitions. Transitions from an excited state to the ground state can occur through the emission of photons which satisfy the following relation:

$$\Delta E = (hc)/\lambda \qquad (3.22)$$

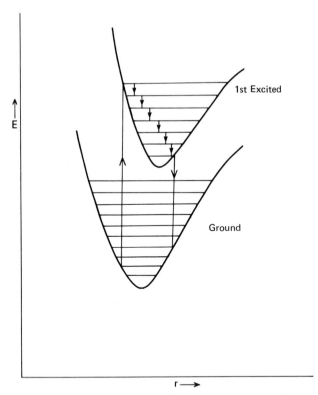

Fig. 3.19. Potential energy curves for the electronic ground state and the first excited state of a diatomic molecule. Transitions must follow the Franck-Condon principle. A transition from a lower vibrational level of the ground state to a higher vibrational level in the first excited state is followed by a rapid vibrational relaxation to lower levels after which emission can take place.

The Stokes shift

The emission spectra, however, do not coincide with the absorption spectra; the peak of the absorption band has a higher energy than the peak of the emission band and therefore, occurs at a lower wavelength. This shift, known as the *Stokes shift*, is the result of the possibility that the excited molecule exchange (thermal) energy with its surroundings by a succession of "downward" transitions between the vibrational substates (Fig. 3.19) before it returns to the ground state. At room temperature the most probable absorption transition (having an energy span corresponding with the peak of the absorption band) originates from the lowest substates in the ground state. Owing to the Franck-Condon principle and the fact that the equilibrium position of the excited state is at a slightly larger internuclear distance, the transition terminates somewhere in the middle of the range of substates in the excited state. This event is then followed by the relaxation of the substates through subsequent intervibrational transitions (giving off heat). The energy of the most probable emission is, therefore, lower than that of the transitions involving absorption. This can be seen from Fig. 3.20, which shows the relative position of a (hypothetical) absorption spectrum and an emission spectrum. The amount of the Stokes shift, since it is related to the time available for the energy relaxation in the excited state, can yield information about the average time spent by a molecule in that state, the lifetime of the excited state.

Internal conversion and intersystem crossing

Potential curves from excited states higher than the first one usually overlap with the potential curve of the first excited state. A number of vibrational states thus are "shared" by two subsequent electronic states and thermal relaxation can (and usually does) occur from higher excited states all the way down to the lower vibrational levels of the first excited state (see Fig.

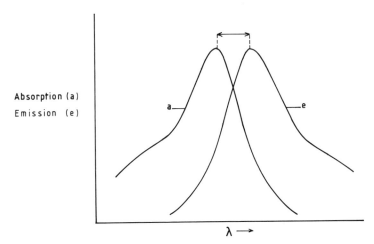

Absorption (a)

Emission (e)

$\lambda \longrightarrow$

Fig. 3.20. A hypothetical absorption spectrum (*a*) and a hypothetical emission spectrum (*e*) showing the Stokes shift.

63

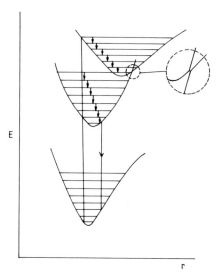

E

r

Fig. 3.21. Potential energy curves for the electronic ground state and two excited states of a diatomic molecule. Excitation of an excited state higher than the first excited state is followed by a rapid vibrational relaxation which extends through commonly shared vibrational levels. Emission thus starts always from the first excited state. The insert shows that the potential curves actually do not overlap but seem to avoid each other.

3.21). This process is known as *internal conversion*. A result of this is that emission always originates from the first excited state, even when the absorption causes a primary excitation into higher states. Even though the absorption spectrum can have more than one band the emission is nearly always correlative with the long wavelength band and represents transitions from the first excited state to the ground state (see Fig. 3.21).

Ordinarily the potential curves for the ground state and the first excited state do not overlap. During a molecular collision, however, the potential curves may become distorted in such a way that the ground state curve temporarily overlaps the first excited state curve. Then internal conversion from the first excited state to the ground state is possible and deexcitation occurs without the emission of photons. This is one of the factors which make the *fluorescence yield* (the ratio between the number of emitted photons and absorbed photons per unit time) smaller than 1.

Other pathways for radiationless deexcitation are transitions to metastable (*e.g.* triplet) states (a process often called *intersystem crossing*); the radiationless transfer of the excitation energy to neighboring molecules and/or the use of the excitation energy for a chemical reaction. Many biological reactions involve one or more of these processes. The alternative pathways for excitation and deexcitation are depicted in Fig. 3.22.

Intrinsic lifetime, absorbance, and fluorescence yield
The intrinsic probabilities for absorption and emission (Einstein transition probabilities) are proportional. The lifetime of an excited state, therefore,

64

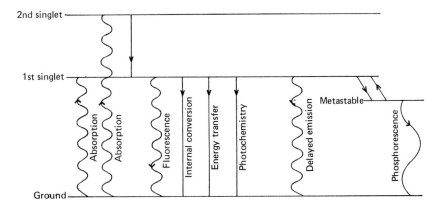

Fig. 3.22. Diagram showing the different modes of deexcitation.

varies inversely as the probability of absorption. Based on this, one can derive a relation between the intrinsic lifetime (τ_0) of an excited singlet state and the integrated absorption coefficient. Since the shapes of most absorption bands are such that the area $\int \epsilon d\nu$ (with ν being the wave number in cm^{-1}) is equal to the halfwidth of the band ($\Delta\nu$) times the value of ϵ at the peak of absorption (ϵ_{max}), the approximate form of

$$1/\tau_0 = 3 \times 10^{-9}\nu^2 \, \Delta\nu \, \epsilon_{max} \tag{3.23}$$

can be used.

The coefficient ϵ in this form is the *molar extinction coefficient* as defined by the *absorbance* (sometimes called *optical density*)

$$A = \log I_0/I = \epsilon C l \tag{3.24}$$

in which C is the concentration in moles per liter, l the optical path length in centimeters, I_0 the intensity of the incident light, and I the intensity of the transmitted light.

Eq. 3.24 is the familiar law of Beer-Lambert. It can be derived as follows: Consider a solution (or a suspension) of absorbing molecules at a concentration of C moles per liter in a vessel with a thickness l (see Fig. 3.23). Incident on the vessel is a light beam with intensity I_0. Imagine a flat thin layer, with a thickness dl, of the solution perpendicular to the direction of light propagation. If the layer is sufficiently thin, the intensity within the layer stays essentially constant. Then the fraction of the light absorbed in the layer, $-dI/I$, is simply proportional to the number of moles (per unit area), Cdl, in the layer, giving the equation

$$-dI/I = \epsilon'C \, dl \tag{3.25}$$

where ϵ', the proportionality constant, is the extinction coefficient. Integration of the differential equation (3.25) gives

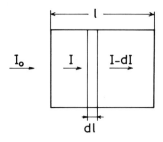

Fig. 3.23. Absorption of light in a sample.

$$\ln (I_0/I) = \epsilon'Cl \qquad (3.26)$$

Converting the natural logarithm in 3.26 to the logarithm base 10 gives equation 3.24 with $\epsilon = \epsilon'/2.303$.

For solutions A is directly proportional to the concentration C and to the pathlength l. The units of the molar extinction coefficient ϵ are the reciprocals of C and l. Thus if C is measured in molar concentration (M) and l in cm, the extinction coefficient is in $M^{-1}cm^{-1}$. It is easy to show that the absorbance is not only proportional to the concentration of a single substance but also additive for a mixture of substances.

The actual lifetime τ of the excited state is less than the intrinsic lifetime τ_0 by a factor φ_f

$$\tau = \varphi_f \tau_0 \qquad (3.27)$$

This factor is the fluorescence yield that we have mentioned above. The actual lifetime, and hence the fluorescence yield, is determined by the rate k of deexcitation; if all modes of deexcitation are independent first-order processes, the rate is

$$k = 1/\tau = k_0 + \sum_i k_i \qquad (3.28)$$

in which the k_i are the rate constants for all deexcitation processes other than fluorescence. Since $\tau_0 = 1/k_0$ and $\varphi_f = k_0/k$, relation 3.27 follows immediately from 3.28.

Absorption difference spectroscopy

If one measures the absorption spectrum of a solution of molecules, what is of interest is the spectrum of the molecule and not that of the solvent. Therefore, it is useful to measure the *difference* in absorbance between a solution of the molecule and pure solvent. Because of the additive property of the absorbance, this is a straightforward procedure. Using a split-beam or a double-beam spectrophotometer (Fig. 3.24) this difference can be determined automatically. The machine measures the logarithm of the ratio of the intensities emerging from the sample (solution) and the reference (solvent). From Eq. 3.24 it follows that the log intensity of the beam emerging from the sample

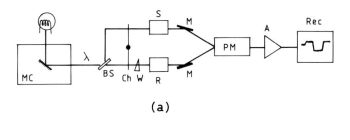

(a)

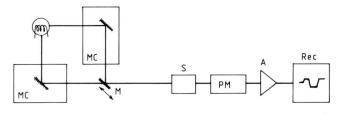

(b)

Fig. 3.24. Schematic representation of methods of difference spectrophotometry. (*a*) A split-beam difference spectrophotometer; light passes a monochromator MC and is partially reflected, partially transmitted by a beam splitter BS; a chopper Ch allows the light of the two beams to pass alternatively through a vessel S containing the sample and a vessel R containing a reference solution; the difference between the two beams (after adjusting the zero level by means of an optical wedge W) is measured by the same photomultiplier PM; its signal is processed by a phase-sensitive amplifier A and recorded. (*b*) A double-beam difference spectrophotometer; light passes two monochromators MC, one set at the measuring wavelength λ_S and the other at an isosbestic (reference) wavelength λ_r; the two beams are allowed to pass alternatively the same vessel containing the sample by means of a vibrating mirror M; the difference between the intensities of the measuring beam λ_S and the reference beam λ_r is measured in the same way as described above.

is

$$\log I_s = \log I_0 - A_s \qquad (3.29a)$$

and that from the reference is

$$\log I_r = \log I_0 - A_r \qquad (3.29b)$$

Therefore, an instrument that detects $\log I_r/I_s$ will measure A_s-A_r.

A very useful extension of this principle is the technique that measures spectral changes in absorbance during a (biochemical or biophysical) process. Very often the spectral changes caused by such processes are small and hidden in areas of large absorbancies. One can use the split beam (Fig. 3.24a) or the double-beam (Fig. 3.24b) method. Due to the fact that both methods are essentially zero-methods, very small spectral changes (in the order of $10^{-3}\%$) can be measured this way. The split-beam method uses a reference vessel containing a sample in which the process is not allowed to take place. In the double-beam method a measuring beam and a reference beam with a different wavelength traverse the same sample vessel; if there is no spectral change of the reacting component at the reference wavelength (which then is called an *isosbestic* wavelength), the apparatus measures the amount of reacting component. The latter method is very useful if one wants to avoid scattered light being measured as a change of absorbance.

Optical Rotatory Dispersion (ORD) and Circular Dichroism (CD)

Nearly all molecules synthesized by biological systems show *optical activity*; if a plane polarized light beam passes an optically active sample the plane of the light is observed to be rotated. This optical activity arises from the lack of symmetry these molecules have, in particular from the presence of asymmetric carbon atoms and from the effect these atoms have on nearby light-absorbing molecules or molecular groups (such light-absorbing molecules and molecular groups are often called *chromophores*). Changes of the conformation of macromolecules, or complexes, containing such chromophores affects strongly the optical activity.

Plane polarized light can be represented by its electric vector E that oscillates sinusoidally in a plane (Fig. 3.25a). After passing through an optically active sample the maximal amplitude is no longer confined to a plane; instead it traces out an ellipse (or actually an elliptical screw). The arcus tangent of the ratio of the minor axis to the major axis is defined as the *ellipticity*, θ, of the transmitted light and is a measure of optical activity. The orientation of the ellipse is another measure of optical activity. The major axis of the

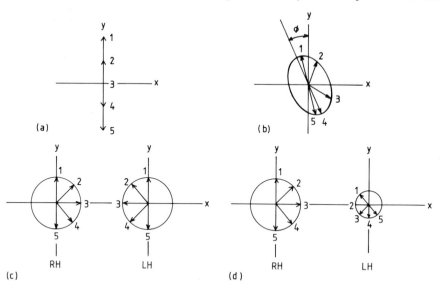

Fig. 3.25. Effect of an optically active sample on incident polarized light. (*a*) Incident linearly polarized light; (*b*) elliptically polarized light produced by passing the incident linearly polarized light through an optically active sample; (*c*) incident right-hand circularly and left-hand circularly polarized light; (*d*) effect of an optically active sample on the two circularly polarized components. The points 1 through 5 correspond to equal (increasing) time intervals. (From C.R. Cantor and P.R. Schimmel, Biophysical Chemistry. Reprinted with the permission of W.H. Freeman and Cy., San Francisco).

68

ellipse is not parallel to the polarization direction of the incident light. If there is little absorption of the light the minor axis of the ellipse is very small in comparison to the major axis. It appears then as if the plane of polarization has been rotated. Thus the orientation of the ellipse can be seen as an optical rotation. The optical rotation as a function of the wavelength is called *optical rotatory dispersion* (ORD).

One can also use circular polarized light instead of plane polarized light (Fig. 3.25c). If left circularly polarized light is combined with right circularly polarized light, the result is simply plane polarized light. It is of interest, however, to examine the effect of an optically active sample on each component separately. It turns out that in such a sample the absorbance of left circularly polarized light, A_l, is different from the absorbance of right circularly polarized light, A_r. Both components are still circularly polarized but the radii of the circles traced out by the electric vector E of each are now different. This phenomenon is called *circularly dichroism* (CD). If one combines the two opposite circularly polarized light waves, the result will be elliptically polarized light because the two amplitudes are different. It can be shown that the ellipticity, θ, is proportional to the difference of absorbance of the two components, $A_l - A_r$. Thus it turns out that CD is equivalent to ellipticity and that both phenomena, CD and ORD, yield redundant information.

The ORD and CD of a sample usually depend strongly on the wavelength of the incident light. One can determine the ORD and the CD over the same wavelength range used for an absorption spectrum thus obtaining ORD respectively CD spectra. The shape of a CD spectrum is often called a Cotton effect. For strongly allowed transitions (such as a $\pi\pi^*$ transition) it is the same as the shape of the absorption peak but it can have a positive or a

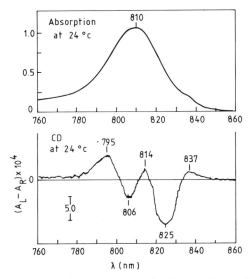

Fig. 3.26. Absorption spectrum and CD spectrum of a chlorophyll-protein complex from a green photosynthetic bacterium (from J.M. Olson et al, Biochim. et Biophys. Acta 292, (1973) 206–217).

69

negative sign. Both, the integrated intensity and the sign of each CD band are sensitive functions of molecular structure. Figure 3.26 shows an absorption spectrum and a CD spectrum in the near infrared region of a chlorophyll-protein complex from a green photo-synthetic bacterium. The spectrum shows that the absorption band is composed of at least three components, one having a negative and two having positive Cotton effects.

3.4 Infrared and Raman spectroscopy

Vibrational energy levels
Transitions between vibrational levels in an electronic state are possible by absorption or emission of the appropriate quanta of energy. For the strict harmonic oscillator such transitions are restricted by the selection rule $\Delta v = \pm 1$, which means that only transitions between adjacent levels are possible. Therefore, if we use Eq. 3.9, the transition energy ΔE is given by

$$\Delta E = h\nu_0 \left[(v + 1) + 1/2 \right] - h\nu_0 (v + 1/2) = h\nu_0 \qquad (3.30)$$

and the frequency of the absorbed or emitted quantum is equal to the characteristic frequency of the oscillator.

A diatomic molecule is not a simple harmonic oscillator. A more realistic potential energy curve is given in Fig. 3.27. The curve approaches an asymptotic level which represents the energy at which the system breaks apart, the so-called dissociation energy. The vibrational energy is given by an equation slightly different from 3.22, but is also determined by the vibrational quantum number v. The levels are not, as in the harmonic oscillator, equidistant from each other, although in the lower levels the deviations from equidistancy are very small. The selection rule $\Delta v = \pm 1$ is not as strict as it is for the harmonic oscillator, although less probable transitions which involve $\Delta v = \pm 2, \pm 3,...$

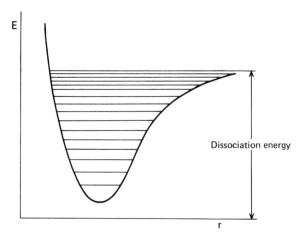

Fig. 3.27. Potential energy curve of a diatomic molecule.

70

are possible and will, according to 3.30, result in the "overtones" approximately $2\nu_0$, $3\nu_0$,....

A big polyatomic molecule can be seen as being made up of many diatomic oscillators. These molecules may, therefore, execute very complex vibrations which cannot be resolved into recognizable single oscillations. A protein molecule with a molecular weight of 10,000, for example, has about 1000 atoms, and the number of coordinates needed to specify the vibrational motion is about 3000. Even the simplest amino acid, glycine, exhibits 24 "diatomic" oscillations. Infrared spectroscopy, by which we can look at vibrational characteristics, therefore, is not very suitable when we want to identify a molecule as a whole. It is very useful, however, as a means of characterizing particular bonds. We can distinguish, in an infrared spectrum, frequencies characteristic of certain bonds if the groups in question are sufficiently isolated from the rest of the molecule and if the frequencies are not too near those of other bonds. Such criteria are satisfied by the end groups of molecules in which the forces holding two atoms together are roughly independent of other atom groups bonded to these groups. If hydrogen is the terminal atom, it can be seen as vibrating against a massive wall.

Stretching and bending vibrations
The vibrations in such end groups can be of the stretching type

$$C \underline{\quad} \overleftrightarrow{H}$$

or the bending type

$$C \underline{\quad} H \Big) .$$

Bending frequencies are of the order of 10^{13} sec^{-1}; stretching frequencies are usually an order of magnitude higher. Characteristic frequencies (in the wave number unit cm^{-1}) of some end groups are given in Table 3.6.

Hydrogen bonds
Infrared absorption spectra of a protein (keratin) and a polypeptide (made of the amino acids phenylalanine and leucine) are shown in Fig. 3.28. We can recognize bands in these spectra as belonging to stretching and bending vibrations of several groups. Upon closer examination we may discover, however, that the stretching frequencies of the NH groups at 3300 and 3200 cm^{-1} are somewhat lower than the value of 3500 cm^{-1} for this group in the gas phase (Table 3.6). This red shift (a shift toward a longer wavelength) of the band is due to the hydrogen bond in which the group is engaged. The effect of hydrogen bonding will be a reduction of the "stiffness of the oscillator" as is indicated in the example of the springs in Fig. 3.29a. For the diatomic group the result will then be a widened potential curve with the energy levels depressed as shown in Fig. 3.29b. The infrared spectral data represent strong experimental evidence for the presence of hydrogen bonds in proteins and polypeptides.

Table 3.6. Some characteristic bond frequencies in cm^{-1} of gases and liquids

Group	Stretching vibration (cm^{-1} + 100 cm^{-1})	Group	Bending vibration (cm^{-1} + 100 cm^{-1})
≡ C–H	3300	≡C–H	700
= C–H	3020	=C (H, H)	1100
– O–H	3680	C–C≡C	300
	(3400)*	N–H	1600
N–H	3500 (3300)*		
C=O	1700		
C=N—	1650		
C=C	1650		
— C≡C—	2050		
= C–C =	900		
P=O	1250–1300		

* Liquid hydrogen bonded.

Raman spectroscopy

An alternative way to gain information about the vibrational states of molecules is Raman spectroscopy. This technique is based on a phenomenon that was described in 1928 by the Indian physicist C.V. Raman. If a transparant sample of molecules is illuminated by an intense beam of light at frequency ν most of it passes through unaffected. A minor fraction of the light is scattered at the same frequency (this is the elastic scattering known als *Rayleigh scattering*) and an even smaller part is scattered at frequencies quite different from ν. The frequency shifts in this case correspond to the vibrational frequencies of the molecules. The phenomenon can be explained as follows:

If a beam of light at frequency ν illuminates a sample the electric field of the light induces an oscillating dipole in the molecules, even if the frequency ν is too low to cause electronic or vibrational excitation. The electric field oscillates as $E(t) = E_0 \cos 2\pi\nu t$. According to Eq. 3.4, the induced dipole moment is

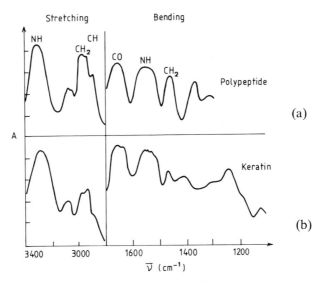

Fig. 3.28. Infrared absorption spectra of (*a*) a synthetic polypeptide, polyphenylanaline-leucine and (*b*) a protein keratin. The peaks due to the NH_2, NH, CH_2, and CO groups are easily detectable. From S.E. Darmon and G.B.B.M. Sutherland, J. Amer. Chem. Soc. **69**, 2074 (1947). Copyright by the American Chemical Society. Reprinted by permission.

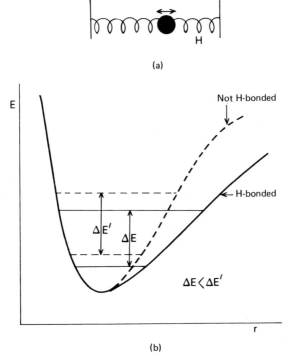

Fig. 3.29. (*a*) A model showing the modification of the vibration due to hydrogen bonding. (*b*) The modification of the potential energy curve due to hydrogen bonding. The curve is stretched and the differences between the vibrational energy levels have become smaller.

73

$$\boldsymbol{\mu}(t) = \alpha(\nu) \, E_0 \cos 2\pi\nu t \,. \tag{3.31}$$

The polarizability α*) depends on the relative movement of the nuclei in respect to the electrons in the molecule. If the molecule is in a vibrational state in which it vibrates with a frequency ν_i, its polarizability contains a part that varies with that frequency. This can be described by writing

$$\alpha(\nu) = \alpha_0(\nu) + \alpha'(\nu) \cos 2\pi\nu_i t \tag{3.32}$$

in which α_0 is the polarizability of the molecule in its equilibrium nuclear configuration and $\alpha'(\nu)$ is the maximum change in polarizability with the motion of the nuclei at vibrational frequency ν_i. Combining the Eqs. 3.31 and 3.32 gives

$$\boldsymbol{\mu}(t) = E_0 \left[\alpha_0(\nu) + \alpha'(\nu) \cos 2\pi\nu_i t\right] \cos 2\pi\nu t$$

$$= E_0\alpha_0(\nu) \cos 2\pi\nu t + E_0\alpha'(\nu) \cos 2\pi\nu_i t \cos 2\pi\nu t \,.$$

Using a well known trigonometric relation yields

$$\boldsymbol{\mu}(t) = E_0 \, \alpha_0(\nu) \cos 2\pi\nu t + 1/2 \, E_0\alpha'(\nu) \cos 2\pi(\nu + \nu_i)t +$$

$$1/2 E_0\alpha'(\nu) \cos 2\pi(\nu - \nu_i)t \,. \tag{3.33}$$

A dipole oscillating at a particular frequency results in the emission of radiation at that frequency. Thus Eq. 3.33 indicates that radiation can be observed at frequencies ν, $\nu + \nu_i$, and $\nu - \nu_i$. The intensity of the radiation (which

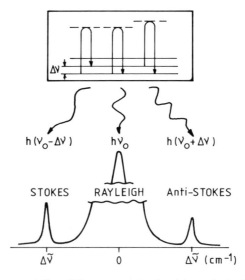

Fig. 3.30. Schematic representation of Raman scattering involving a single vibrational mode.

---

*The polarizability, in fact is a tensor. For our discussion here, however, we can consider it as a number.

is proportional to $|\mu|^2$) is very low compared to the intensity of the incident light and is emitted in all directions. The major part of the intensity is emitted at frequency ν and appears as Rayleigh scattering. This is described by the first term of Eq. 3.33. The other two terms of the equation predict radiation in bands at $\nu + \nu_i$ and $\nu - \nu_i$.* The lower energy band is called the *Stokes band*, the higher energy band the *anti-Stokes band*.

For a single vibrational system the phenomenon can be illustrated as in Fig. 3.30. A molecule in the ground state interacts with an incident photon with energy $h\nu$ and is excited in a metastable, very short-lived *virtual* state. Upon emitting a photon of energy $h(\nu - \nu_i)$ it returns to the ground state but to the first excited vibrational level rather than to the ground vibrational level. This describes Stokes scattering. Anti-Stokes scattering occurs if the incident photon interacts with molecules which populate vibrational levels above the ground state level. Since the number of molecules populating higher vibrational levels is small, anti-Stokes bands usually are very weak.

For small molecules, the Raman effect tends to complement infrared spectroscopy. Transitions that produce Raman bands must have a polarizability that changes with nuclear motion, rather than a permanent dipole that changes. In the latter case infrared spectroscopy produces the better information about vibrational states. For large asymmetric molecules the two methods give

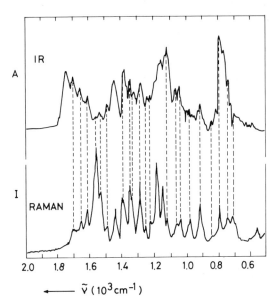

Fig. 3.31. Infrared spectrum (*a*) and Raman spectrum (*b*) of partly aggregated chlorophyll *a* in dry hexane (from E. Höxtermann, Thesis 1985, Humboldt University, Berlin DDR).

---

*This is not in contradiction with the principle that no radiation can be emitted with energy higher than the energy of the exciting light. We can view the Raman effect as a two-photon process; the molecule takes up two photons with energy $h\nu$ to emit one photon with energy $h(\nu - \nu_i)$ and one photon with energy $h(\nu + \nu_i)$. The overall change in energy then is zero.

75

essentially the same set of bands (cf. Fig. 3.31). A major advantage of Raman spectroscopy, applied to biological systems, is that water has a weak Raman spectrum. Therefore, biological samples can be measured in aqueous solutions and in cells and cell fractions.

Raman spectroscopy has been applied extensively to biological molecules only since lasers, combining large intensities with high spectral purity, became available. Moreover, the many different (often tuneable) frequencies of the present-day lasers allow a powerful variant of the technique: If the frequency of the light used to produce the Raman spectrum is in, or in the neighborhood of an absorption band of a chromophore in the sample, the coupling of electronic and vibrational transitions in the chromophore results in a substantial enhancement of the Raman bands localized on the chromophore. This variant, which is known as *resonance Raman spectroscopy*, allows the selective study of the vibrational spectrum of an individual chromophore even in the presence of many other disturbing vibrational bands.

3.5 Magnetic resonances

Magnetic resonance spectroscopy
Many atomic nuclei have an intrinsic spin angular momentum, just like electrons have. Since nuclei have positive charges, a nucleus with spin must also have a magnetic moment, which is always parallel to the spin vector. For both spinning nuclei and electrons, the orientation of these vectors is quantized; only a discrete number of orientations is possible. In a magnetic field these differences in orientations lead to different interaction energies with the field and therefore to different energy levels. Transitions between these energy levels become possible and can be observed as resonance absorption of electromagnetic radiation of the appropriate wavelength. Such phenomena are the bases of *Nuclear Magnetic Resonance* (NMR) and *Electron Paramagnetic Resonance* (EPR)* spectroscopy. With both techniques the absorption of energy associated with transitions of nuclei (in NMR) or electrons (in EPR) between adjacent energy levels is monitored. The energy is measured as a function of the magnetic field strength or of the frequency of electromagnetic energy applied by an oscillator.

Nuclear Magnetic Resonance
The value of nuclear magnetic resonance spectroscopy lies in the fact that individual nuclei of a given kind within the same molecule (such as hydrogen nuclei at chemically distinct positions) normally absorb at distinct positions in the spectrum. The absorption characteristics of a nucleus are extremely sensitive to the local magnetic environment. Therefore, hydrogen nuclei within the same molecule can often be resolved. Small molecules and molecular groups

---

*In the literature very often the term *Electron Spin Resonance* (ESR) is used.

within larger ones thus can be "fingerprinted" and NMR spectra of molecules with only slightly different structures can be distinguished. Because of these features, the technique can be used for studying specific loci in biological macromolecules under various conditions. Also quantitative information can be obtained on the exchange rates of molecules which exchange between different magnetic environments.

An accurate and complete discussion of the phenomenon of nuclear magnetic resonance cannot be given in this book. Entire books are devoted to the subject alone. In the following we shall discuss some of the general principles. In order to do that, we need first to describe the interaction of a spinning nucleus, having an angular momentum, with an external magnetic field. Quantum mechanics tells us that the magnitude of the angular momentum of such a spinning nucleus is given by

$$| L | = \hbar \sqrt{I(I+1)}. \tag{3.34}$$

in which I is the spin quantum number. I is zero for nuclei with even mass numbers and even atomic numbers.* These nuclei have no spin. The isotopes ^{12}C and ^{16}O are examples. Half-integral values of I correspond to nuclei with odd mass numbers (such as ^{1}H and ^{31}P, both with $I=1/2$) and integral values of I to nuclei with even mass numbers and odd atomic numbers (such as ^{14}N with $I = 1$). The relation between the magnetic moment m of a nucleus with $I \neq 0$ and its angular momentum L is given by

$$m = \gamma L \tag{3.35}$$

where γ is the *nuclear magnetogyric ratio*. Originally it was thought that, analogous to an electron the magnitude of the magnetic moment of a proton (a hydrogen nucleus) would be one *nuclear magneton* (equal to $e\hbar/2M_p$ in which M_p the mass of the proton). It turned out, however, that a proton has a magnetic moment that is a factor $g_n = 2.792$ times larger. The factor g_n is called the *nuclear g-factor*. The magnetogyric ratio, therefore, can be written as $\gamma = g_n \mu_n$ in which μ_n is the nuclear magneton.

The orientation of the magnetic moment with respect to the z direction (determined, for instance, by the direction of an external magnetic field) is quantized; the component of the magnetic moment in the z direction can have the values

$$m_z = m_I \gamma \hbar \tag{3.36}$$

in which $m_I = I, I-1, I-2,....I-2I$. Thus, for a proton with spin 1/2 only the two orientations $m_I = +1/2$, or $-1/2$ are possible. If an external magnetic field B is applied to the system the energy E associated with the interaction between B and a magnetic moment m is

---

*The mass number refers to the total number of protons and neutrons in the nucleus; the atomic number to the number of protons.

$$E = -\boldsymbol{m}.\boldsymbol{B}. \qquad (3.37)$$

The interaction tends to align the magnetic moments with respect to the direction of the external magnetic field. The only two alignments which are possible are those with $m_I = +1/2$ (parallel to the field direction) and with $m_I = -1/2$ (antiparallel to the field direction), respectively with energies $+(1/2)\gamma\hbar B$ and $-(1/2)\gamma\hbar B$. Transitions between these two levels occur when the field strength B and the incident electromagnetic radiation match the energy difference $\Delta E = \gamma\hbar B$ as shown in Fig. 3.32a.

Considering a collection of spinning nuclei, each having a magnetic moment $\boldsymbol{m}$, we can use a classical picture to describe the magnetic resonance pheno-menon. In the absence of an external field these nuclei and their magnetic moments are randomly oriented. Therefore, the net magnetization $\boldsymbol{M}$, which is equal to the vector sum of the magnetic moments is zero. When the external magnetic field is turned on, the nuclei will show a preference to align with the field direction and start to perform a precession about the axis of the field, just like a top that precesses about its rotation axis. The angular velocity of the precession, ω, is called the *Larmor frequency*. By using the classical equations describing such a precession, it is easy to show that the Larmor frequency is given by

$$\boldsymbol{\omega} = -\gamma\boldsymbol{B}. \qquad (3.38)$$

A net magnetization $\boldsymbol{M}$ occurs, the orientation of which (returning to quantum mechanics) is quantized. Its components in the direction of the field (taken as the z direction) are given by the values of a quantum number M_I (Fig. 3.32b). If one applies an oscillating magnetic field with oscillating frequency ω, energy is absorbed from the field. In other words, quanta of electromagnetic

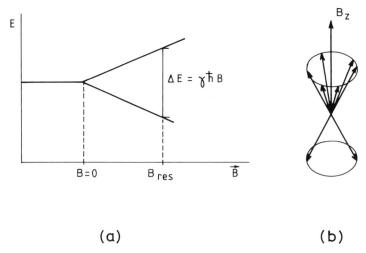

(a) (b)

Fig. 3.32. (*a*) Splitting of spin states of a nucleus with I=1/2; (*b*) Classical description of precessing magnetic moments with I=1/2. The net magnetization M is along the z-axis aligned with the field.

radiation are absorbed as the quantum number M_I increases by one unit. The resonance effect measures the net energy absorption, that is the difference of the energy absorbed when the system goes from the lower state to the upper state and the energy emitted as the system goes from the upper state to the lower state. Since in equilibrium the lower state is slightly more densely populated than the upper state there is a net absorption of energy.

A schematic view of an NMR experimental setup is given in Fig. 3.33. The magnetic field B is supplied by a permanent magnet. Resonance absorption is achieved by changing the frequency of the electromagnetic field (frequency of the radio transmitter) or, as is done more often, by varying the magnetic field B using a secondary variable field known as the *sweep field*. In the latter case the frequency of the electromagnetic field, provided by the transmitter coil, is kept constant. Resonance absorption is measured by the receiver coil. Both transmitter coil and receiver coil are perpendicular to the magnetic field.

Relaxation

No NMR experiment would be possible if there would be no way to restore the thermal equilibrium which keeps the lower states more populated than the upper ones. The system can return to the lower state, not only by emission but also by various radiationless mechanisms, which are called *relaxation processes*. There are two kinds of relaxation processes. *Longitudinal relaxation* causes the return to the equilibrium value of the nuclear magnetization in the z-direction (the direction of the external field). The z-component of the magnetization, M_z, returns to its equilibrium value M_{0z} in a simple first-order fashion, so that

$$M_z = M_{0z}(1 - e^{-t/\tau_1}) . \tag{3.39}$$

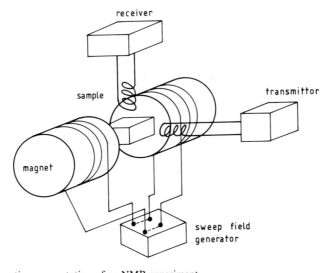

Fig. 3.33. Schematic representation of an NMR experiment.

The time constant τ_1 is the *longitudinal relaxation time*. The process is due to the interaction of the magnetic moments with various fluctuating local magnetic fields in the environment of the oriented nuclei and, therefore, is also called *spin-lattice relaxation*. The addition of paramagnetic ions can greatly shorten τ_1.

The other relaxation process is the *tranverse relaxation*. Normally, nuclei precessing about the direction of the external field do this randomly; although they all precess with the same velocity ω they have no phase relation among themselves. As a result, the magnetization M has components in the z-direction but no components in the xy-directions. If the nuclei are brought in phase (by the variable field that can be seen as an additional magnetic field that rotates in the xy plane with the same angular velocity ω) a net component in the xy plane is generated. Any disturbing field that destroys the phase coherence causes relaxation of this component. A process that provides for such a disturbing field is *spin-spin* relaxation, in which a nucleus in a higher spin state exchanges spin with a neighboring nucleus. Also this relaxation process is first-order with a transverse relaxation time τ_2.

Chemical shifts

The local environment of a spinning nucleus has a small but measurable effect on the interaction of the nucleus and the external field. This is the reason why NMR is useful to begin with. If there were no such effects, protons in various compounds or at different positions of the same compound, would resonate at the same frequency and therefore would be indistinguishable. The effect of the local environment is due largely to the fact that when a field is applied to an atom, circulating electron currents are set up that generate a small magnetic field opposite to the applied field. This shielding effect amount to only some 10 parts per million of the external field, but the sensitivity of the NMR measurements, especially in the case of protons is so great that the effect is readily measured. The result of the effect is called *chemical shift*. It is expressed in terms of a chemical shift parameter δ, relative to the chemical shift of some standard substance

$$\delta = \frac{B_{ref} - B_{samp}}{B_{ref}} \times 10^6$$

or, using Eq. 3.38,

$$\delta = \frac{\nu_{ref} - \nu_{samp}}{\nu_{ref}} \times 10^6 \qquad (3.40)$$

in which $\nu = \omega/2\pi$. The subscripts ref and samp refer to the standard substance and the sample respectively. The factor 10^6 comes in because chemical shifts usually are expressed in parts per million. For protons in aqueous solution the standard substance most widely used is 2,2-dimethyl-2-silapentane-5-

80

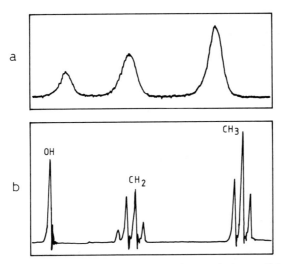

Fig. 3.34. NMR spectrum of ethanol. (*a*) Spectrum at low resolution at 40 MHz; (*b*) spectrum in the presence of a trace of acid; the spin-spin splitting due to the interaction between CH_2 and CH_3 protons is observed but the OH proton exchanges so rapidly between the molecules that no splitting of its resonance peak is observable. (After W.J. Moore, Physical Chemistry, 4th edition (1972), p. 816. Reprinted with the permission of Prentice Hall, Englewood Cliffs, NJ).

sulfonate (DSS) whereas tetramethylsilanate (TMS) is commonly used for nonaqueous solvents.

Spin-spin splitting

Figure 3.34a shows an NMR spectrum of ethanol. The three peaks whose areas are in the ratio 1:2:3, reflect the resonances of the protons of, respectively the OH group, the CH_2 group and the CH_3 group in the molecule HO–CH_2–CH_3. They show up in different positions of the spectrum because their magnetic environment is different. Figure 3.34b shows the same spectrum but measured at higher resolution. Both the CH_2 and the CH_3 peaks are split, the CH_2 peak into four and the CH_3 peak into three lines. The spacing *J* between adjacent lines is the same for each group and is called the *coupling constant*. This effect is not due to chemical shift but to the interaction between adjacent nuclear spins. It is therefore called *spin-spin splitting*. The number of lines and their relative intensities can be predicted by considering the possible orientations of the magnetic moments of the protons in each set. On the left hand side in Fig. 3.35 the relative orientations of the proton spins in the CH_2 group are shown. There are four different orientations but two of these have the same energy. The protons in the adjacent CH_3 group therefore are in three distinct nuclear-spin environments, so that their resonance band is split in three lines about in the ratio of 1:2:1. On the other hand, there are eight different relative orientations of the proton spins in the CH_3, six of them divided over the two distinct energy levels. The protons of the neighboring CH_2 group therefore are in four distinct environments resulting

81

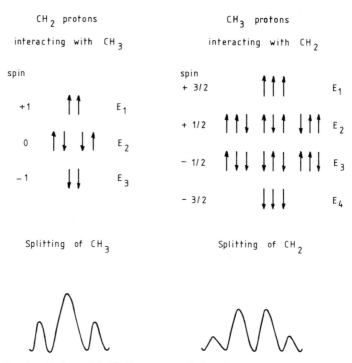

CH$_2$ protons interacting with CH$_3$

CH$_3$ protons interacting with CH$_2$

spin
+1 E$_1$
0 E$_2$
−1 E$_3$

spin
+ 3/2 E$_1$
+ 1/2 E$_2$
− 1/2 E$_3$
− 3/2 E$_4$

Splitting of CH$_3$

Splitting of CH$_2$

Fig. 3.35. Spin-spin splittings of the NMR spectrum of ethanol.

in a fourfold splitting of the resonance band in an approximate 1:3:3:1 ratio.

In a modern development, both spin-spin coupling and chemical shifts are measured in one experiment and plotted in a 2-dimensional spectrum. The technique makes use of Fourier transforms and allows resolutions that otherwise are not attainable. For a description of this technique we refer to the literature (see the *Bibliography*).

Electron Paramagnetic Resonance

Also spinning electrons have magnetic moments that interact with external magnetic fields. In most biological macromolecules, however, the electrons occur in pairs which, according to the Pauli principle, have opposite spins. Their magnetic moments cancel and therefore no resonance signal is measured. Only those molecules that have one or more unpaired (paramagnetic) electrons yield an EPR spectrum. This apparent limitation turns out to be an advantage in some biophysical applications; paramagnetic species that are formed in some biological processes serve as internal probes that can be studied without interfering effects of resonance bands from the many other groups in the environment. If no internal paramagnetic species is present, one can attach a site-specific so-called spin label (a small molecule containing an unpaired electron) to the molecule in order to investigate in detail structure-function relationships at and near the specific attachment site.

The discussion about NMR in the previous paragraphs applies equally well

82

to EPR. Also an unpaired electron in an external magnetic field can be seen as precessing about the direction of the field with a component of its spin angular momentum either parallel (if the spin quantum number $m_s = +1/2$) or antiparallel (if $m_s = -1/2$) to the field (z) direction. When the frequency of an applied electromagnetic field matches the precessing frequency ω, transitions between the two states corresponding to the two components are induced and energy is absorbed. The precessing frequency is $\omega = -\gamma B$ in which γ is the magnetogyric ratio. For the electron this parameter is given by $\gamma = -g\mu_B/\hbar$ in which μ_B is the Bohr magneton (Eq. 3.8). The minus sign comes in because the electron is negatively charged.

For an unpaired free electron the g-factor is 2.0023. This means that $\gamma = -1.7 \times 10^{11}$ Tesla^{-1}sec^{-1} which is in absolute value about 10^3 times larger than the absolute value of γ for the commonly studied nuclei. The resonance frequency, therefore is in the GHz ($=10^9$ Hz), hence microwave, range rather than in the radio frequency range in which NMR operates. It is customary

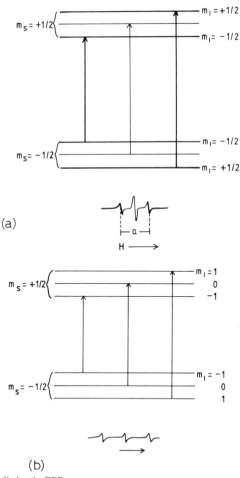

(a)

(b)

Fig. 3.36. Hyperfine splitting in EPR spectra.

in EPR to use a fixed frequency for the electromagnetic source and to vary the magnetic field until the value B_{res} is reached at which resonance occurs. Since information extracted from EPR spectra often is obtained by detailed line shape analysis and accurate resolution of closely spaced lines (see below), EPR spectra usually are registered as first derivative spectra (see Fig. 3.36).

g-factor and hyperfine splitting
The energy of an unpaired electron in a magnetic field depends on the parameter g. As mentioned before, the g-factor of a free unpaired electron is 2.0023. However, local magnetic fields induced by the external field can cause variations of the value of g. Also, the value of g generally varies with respect to the orientation of the molecule relative to the magnetic field. This can be useful, because it allows recognition of the resonance band as belonging to a special environment. Unpaired electrons belonging to some metal ions, for instance, have g-values that differ substantially from the g-value of a free electron.

Another useful feature of EPR is the splitting of an EPR band as a result of the effect of local permanent fields. The magnetic environment of an electron not only consists of the externally applied magnetic field but also of local permanent fields from the magnetic moments of nuclei. These local fields combine with the external field to give an effective field B_{eff}. If an electron is essentially localized around one nucleus with a spin quantum number I the contribution of the local field can assume $2I+1$ values, corresponding to the $2I+1$ values of the quantum number m_I. Therefore, EPR occurs at m_I values B_{res}, of the external field. This is expressed as

$$B_{res} = B^{\circ}{}_{ref} - am_I \quad \text{for } m_I = -I, -I+1, \ldots, I-1, I \quad (3.41)$$

in which a is a constant such that aI is the magnitude of the local field, and $B^{\circ}{}_{res}$ is the value of the field at resonance if $a = 0$. The parameter a is called the *hyperfine splitting constant*. Figure 3.36 shows the transitions and the resulting EPR (derivative) spectra for an electron near a nucleus with spin $I = 1/2$ (a) and a nucleus with $I = 1$ (b).

Spin-label EPR
As mentioned earlier, most biological macromolecules do not give an intrinsic EPR signal. Such molecules nevertheless can be studied by EPR if a site-specific paramagnetic probe (a so-called *spin label*) is attached to the molecule. A spin label is a small molecule that contains an unpaired electron. For many such studies derivatives of the nitroxide radical are used. Care must be taken that the attachment of the probe disturbs the molecule as little as possible. The perturbing effect can be measured, for instance by checking the biological activity with and without the probe attached. If appropriately inserted, the signal from the spin label reflects information about the structure, the state, and the dynamics of the native system. Such information is available because of the sensitivity of the g-factor and the hyperfine splitting constant to the orientation of the radical with respect to the field as well as to the polarity

of the environment of the radical and to motion.

3.6 Size and shape of biological macromolecules

Most biological macromolecules have molecular weights well over 10,000. Since these molecules are in some way or another engaged in the multitude of reactions in a living cell and since the progress and control of these reactions depend on their arrangement and structure, a knowledge of their sizes, shapes, and detailed structure is imperative for understanding their functions and the processes in which they are involved. A number of techniques are used to determine these structures. The most detailed information about the fine structure can be obtained by X-ray diffraction, a technique which is described in section 3.7. Information about size and shape of macromolecules can be obtained by examining the way in which size and shape affect their ability to move in a fluid medium or to look at the way in which size and shape affect phenomena as light scattering and osmotic behavior. Estimates of molecular weight and shape can be obtained this way albeit at a lower level of detail.

Osmotic pressure
Many of these techniques involve the observation of deviations due to large size from the idealized behavior of molecules, molecular groups, or particles under certain conditions; osmotic pressure is here given as an example.

For dilute solutions the osmotic pressure Π obeys the well-known Van 't Hoff relation

$$\Pi V = nRT \text{ or } \Pi = CRT \tag{3.42}$$

in which V is the volume, n the number of moles, C the concentration, R the gas constant, and T the temperature in absolute units. Experience shows that this relation accurately describes situations for which the concentration does not exceed 1%. However, for far more dilute solutions of large molecules (such as rubber in benzene) extensive deviations from the Van 't Hoff relation are observed. The questions which then arose were (1) how can the molecular weight at which failure of the Van 't Hoff law begins be determined and (2) how can the data on osmotic pressure be used to determine the size and shape of large molecules? McMillan and Mayer (1945) showed that the osmotic pressure Π of solutions of nonelectrolytes can be represented by a power series of C_g, the concentration in grams per cubic centimeter:

$$\Pi = \frac{RT}{M} C_g + a_2 C_g^2 + a_3 C_g^3 + \ldots \tag{3.43}$$

in which M is the molecular weight of the solvent molecules and a_2, a_3,... are coefficients which are independent of C_g. For fairly dilute solutions the third- and higher-order terms can be neglected so that 3.43 can be re-written as

$$\frac{\Pi}{C} = \frac{RT}{M} + a_2 C . \tag{3.44}$$

Thus, when measurements of the osmotic pressure Π at different values of the concentration C are made, a plot of Π/C as a function of C should yield straight lines at the lower values of C. The molecular weight M can then be determined from the intercept $C \to 0$.

Light scattering
Rayleigh worked out a theory of light scattering by isotropic particles whose dimensions were small compared with the wavelength of the incident light. According to this theory the relation between the intensity I_θ of the light scattered at an angle θ, and observed at a distance R_s from the scattering volume, and the intensity I_0 of the incident light is

$$I_\theta R_s^2 / I_0 = R_\theta (1 + \cos^2\theta) \tag{3.45}$$

in which R_θ, the *Rayleigh coefficient* or the *Rayleigh ratio*, is given by

$$R_\theta = 8\pi^4 \alpha^2 / \lambda^4 \tag{3.46}$$

in which α is the polarizability of the particle.

When the dimensions of the particle are no longer small, as compared to the wavelength of the light, the theory becomes more complicated because interference effects of the light scattered at different parts of the particle have to be taken into account. In an extension of the Rayleigh theory, Debye, in 1947, showed that for dilute solutions of macromolecules the Rayleigh coefficient is given by

$$R_\theta = K C_g M$$

in which C_g is the concentration in grams per cubic centimeter, M is the molecular weight, and K is a function of the concentration, the refractive indices of the solvent and the solution, and the wavelength. For more concentrated solutions the relation

$$K \frac{C_g}{R_\theta} = \frac{1}{M} + \frac{2aC_g}{M^2}$$

gives a more accurate value for M. The influence of the size of the molecules can be accounted for by a particle scattering factor P(θ), thus yielding

$$K \frac{C_g}{R_\theta} = \frac{1}{MP(\theta)} + \frac{2aC_g}{M^2} \tag{3.47}$$

The scattering data can be obtained from measurements of the scattered

86

light at different values of the scattering angle θ and the mass concentration C_g, extrapolated to zero angle and zero concentration. Thus, from Eqs. 3.45 and 3.47 values for M can be obtained. Measurements of light scattering are also used to determine the diffusion constant (see below).

Friction and viscosity
When large molecules move through a liquid with a velocity v they undergo a frictional force

$$F = -fv . \qquad (3.48)$$

The frictional coefficient f is related to the molecular size and shape, and to the viscosity η of the liquid. If we can determine f, and the constituents and concentration of the molecules are known, we can calculate the molecular weight, provided that the grammolecular volume can be determined from the shape parameters. The latter is rigorously possible only for spherical molecules for which Stoke's law

$$f = 6\pi\eta a \qquad (3.49)$$

a being the radius of the molecule, is valid, and for ellipsoid molecules for which a similar relation can be derived. For more complicated shapes, approximations have to be applied in order to arrive at an estimate of the grammolecular volume.

Diffusion
In order to determine f we must have the molecules move through a medium. This can be accomplished by making use of the process of diffusion in which the driving force for the movement is provided by a concentration gradient, or by sedimentation in which the driving force is gravity or centrifugal force. Diffusion is a direct result of the random motion of the molecules. If there is a region of high concentration in a vessel, the net random movement will be out of that region and into regions of lower concentration. Fick's law states that if dn/dt is the number of molecules that in one dimension pass through a cross section A per unit time

$$dn/dt = -DA \, \partial c/\partial x \qquad (3.50)$$

in which $\delta c/\delta x$ is the (one-dimensional) concentration gradient and D is the *diffusion constant*. This diffusion constant is related to the friction coefficient

$$D = kT/f \qquad (3.51)$$

in which k is the Boltzmann constant and T is the absolute temperature. Table 3.7 (see below) lists values of D for some protein molecules.

By examining how concentrations (which can be measured, for instance, by optical methods) are changed with time by concentration gradients, one could determine the diffusion constant. Figure 3.37 is a schematic representation of a method which could be used for this purpose. Two containers, one filled with a solution at the initial concentration $c = c_0$ and the other with the

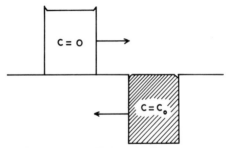

Fig. 3.37. Schematic representation of a diffusion experiment. Two containers, one containing a solvent and the other containing a solution with concentration c_0, are moved so as to bring the two solutions into contact at $t = 0$, thus forming a sharp initial boundary.

solvent ($c = 0$), can be moved so as to bring the solutions in contact with each other at time $t = 0$. Since the number of diffusing molecules in such an experiment is conserved, the increase in concentration within a volume of cross section A and length dx is the excess of material diffusing into that volume over that diffusing out. This leads to the diffusion equation

$$\frac{\partial c}{\partial t} = D \frac{\partial^2 c}{\partial x^2}.$$ (3.52)

The boundary conditions (for $t = 0$), determined by this experiment, are $c = 0$ for $x < 0$ and $c = c_0$ for $x > 0$. A solution is the probability integral

$$c = 1/2\, c_0 \left[1 + \frac{2}{\sqrt{\pi}} \int_0^{x/(4Dt)^{1/2}} e^{-\xi^2}\, d\xi \right]$$ (3.53)

the values of which are widely tabulated. From 3.53 it follows, for the concentration gradient, that

$$\frac{\partial c}{\partial x} = \frac{c_0}{\sqrt{4\pi Dt}} e^{-x^2/4Dt}.$$ (3.54)

Figure 3.38 shows the concentration (a) and the concentration gradient (b) as functions of x at different values of t.

The diffusion could be followed by optical methods, and the diffusion constant D could be calculated from the tables of the probability integral. However, recent developments in light scattering spectroscopy (photon correlation spectroscopy; see, for instance Chen and Yip, 1974) which came about with the development of laser technology have made the determination of the diffusion constant much more easy. The linewidth Γ of scattered (originally coherent) laser light is equal to

$$\Gamma = 2D|\mathbf{k}|^2$$ (3.55)

in which $\mathbf{k}$ is the scattering vector. This vector is given by $\mathbf{k} = 2\mathbf{k}_0\sin(1/2\,\theta)$ in which $\mathbf{k}_0$ is the wave vector of the incident laser light and θ is the scattering angle. Thus, by measuring the linewidth of the scattered laser light at different

88

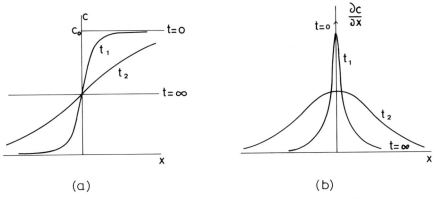

Fig. 3.38. (a) The concentration and (b) the concentration gradient of a diffusing solution as a function of distance at different times after the time t_0 at which the solution came into contact with solvent.

scattering angles and plotting Γ against $|\mathbf{k}|^2$, a direct and unique determination of D is possible. There is no longer a need to establish a concentration gradient in order to initiate diffusion or to avoid complications from mechanical disturbances, gravitation, temperature gradients, and electrical charges on the molecules or particles.

Sedimentation
The masses of particles or large molecules are usually determined by observing their motion under the influence of outside forces such as gravity or centrifugal force. In a gravitational field the terminal velocity of a particle is determined by the balance between the force of gravitation and the frictional resistance of the fluid medium to the motion of the particle. The force of gravitation, which is the difference between the weight of the particle and the buoyant force due to the displacement of the fluid medium, is $mg - \rho_0 Vg$, where m is the mass of the particle, V its volume and ρ_0 the density of the fluid medium. Since $V = m/\rho$, where ρ is the density of the particle, the force of gravitation can be written as $mg(1 - \rho_0/\rho)$. When this force balances with the frictional force fv, the uniform velocity v at which the particle sediments will be

$$v = \frac{mg}{f}\left(1 - \frac{\rho_0}{\rho}\right). \tag{3.56}$$

In general, the weakness of the gravitational field restricts the observation of the sedimentation of particles in such a field to quite massive particles.

The development of centrifugal techniques that can generate "gravitational fields" up to $300{,}000g$ makes this method suitable for large molecules such as proteins, as well as smaller molecules. In this case g in Eq. 3.56 has to be replaced by $\omega^2 r$, in which ω is the angular velocity of the centrifuge and r is the distance between the particle and the center of rotation. The *sedimentation constant*

$$s = v/\omega^2 r \tag{3.57}$$

89

is a characteristic constant for a given molecular species in a given solvent. It is expressed in *Svedberg units* after The Svedberg, who was instrumental in the development of the technique. One Svedberg unit is equal to 10^{-13} sec. Combining 3.56 and 3.57 then yields

$$s = \frac{m}{f} \left(1 - \frac{\rho_0}{\rho}\right) \tag{3.58}$$

Since f is determined only for spherical and ellipsoid particles we would like to eliminate it, for instance by an additional measurement. When the diffusion constant is known, we can make use of relation 3.51 giving, for the mass m,

$$m = \frac{skT}{D[1 - (\rho_0/\rho)]} \tag{3.59}$$

and for the molecular weight,

$$M = \frac{sRT}{D[1 - (\rho_0/\rho)]} \tag{3.60}$$

in which R is the gas constant. Table 3.7 lists the sedimentation constant, diffusion constant, molecular weight, and the inverse of the density (or specific volume) of some molecules at 20°C.

Sedimentation rate and equilibrium
The sedimentation rate of a molecule can be determined by rotating a suspension (solution) of the molecule in an analytical centrifuge (Fig. 3.39) and observing the moving boundary. By making use of a special optical setup

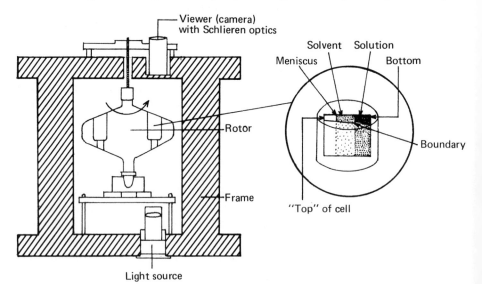

Fig. 3.39. Schematic representation of an ultracentrifuge and the centrifuge cell.

90

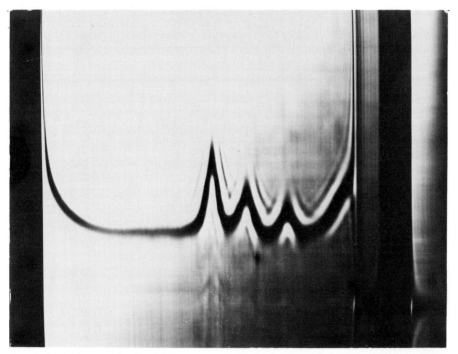

Fig. 3.40. A photograph of a Schlieren pattern of a suspension of ribosomes from the colon bacterium *E. coli.* The three peaks correspond to (from right to left) the sedimentation rates of the two subunits, 30 S and 50 S, and that of the whole ribosome, 70 S.

known as *Schlieren optics* one can record the boundary in terms of a concentration gradient. One can thus obtain photographs of typical Schlieren patterns such as those shown in Fig. 3.40 for a preparation of ribosomes from the colon bacterium *Escherichia coli.*

At equilibrium the rate at which the particles (molecules) move from the rotation axis just equals the rate at which they move in the opposite direction as a result of the concentration gradient. The rate at which the particles sediment (per unit area) is equal to the velocity of sedimentation times the concentration:

$$\frac{dn}{dt} = vc = c\frac{M\omega^2 r}{RT} D(1 - \rho_0/\rho) . \qquad (3.61)$$

The rate at which the particles diffuse back (per unit area) is given by Fick's law (eq. 3.50). At equilibrium the two rates are equal and we obtain

$$\frac{dc}{c} = \frac{M\omega^2}{RT} D(1 - \rho_0/\rho)r \, dr . \qquad (3.62)$$

Integration of the differential equation 3.62 between r_1 and r_2 yields

$$M = \frac{2RT \ln(c_2/c_1)}{[1 - (\rho_0/\rho)]\omega^2(r_2{}^2 - r_1{}^2)} . \qquad (3.63)$$

91

Table 3.7. Characteristic constants of some proteins molecules at 20°C.*

Proteins	$1/\rho(cm^3g^{-1})$	$s(10^{-13}s)$	D (cm^2s^{-1}) $\times10^7$	M $\times10^{-3}$
Myoglobin				
(beef heart)	0.741	2.04	11.3	16.9
Hemoglobin				
(horse)	0.749	4.41	6.3	68
Hemoglobin (man)	0.749	4.48	6.9	63
Hemocyanin				
(octopus)	0.740	49.3	1.65	2800
Serum albumin				
(horse)	0.748	4.46	6.1	70
Serum albumin				
(man)	0.736	4.67	5.9	72
Serum globulin				
(man)	0.718	7.12	4.0	153
Lysozyme				
(egg yolk)	(0.75)	1.9	11.2	16.4
Edestin	0.744	12.8	3.18	381
Urease				
(jack bean)	0.73	18.6	3.46	480
Pepsin (pig)	(0.750)	3.3	9.0	35.5
Insulin (beef)	(0.749)	3.58	7.53	46
Botulinus toxin A	0.755	17.3	2.10	810
Tobacco mosaic				
virus	0.73	185	0.53	31400

* From W. Moore (1972), *Physical Chemistry*, 4th Edn., p. 939. Reprinted by permission of Prentice Hall, Inc., Englewood Cliffs, New Jersey, U.S.A.

This expression 3.63 for the molecular weight is independent of the size and shape of the particles (molecules). The sedimentation equilibrium method, therefore, does not need an independent measurement of D in order to fix the molecular weight. The time required to reach the equilibrium condition, however, is so long that the method is not practical for substances having a molecular weight greater than 5000. A modification of the method makes use of the fact that at the meniscus and at the bottom of the cell there cannot be any net flux. Since the equilibrium condition states that there is no net flux at all times across any plane in the solution, the measurement of the concentrations at the plane of the meniscus and at the plane of the bottom of the cell shortly after the centrifuge is brought to speed could be used to give the equilibrium values.

Density gradient
If a solution of a substance of low molecular weight (e.g. sucrose) is centrifuged, there will be a density gradient across the cell at equilibrium. If we add a substance of high molecular weight to the cell it should float in this solution of varying density at the particular position at which its buoyant density equals the density of the solution. If the substance is made up of various fractions

of different molecular weight, each fraction should separate out in a band at a particular plane of the solution. This method of separation is called the *density-gradient* method. It is widely used in biochemical and biophysical research.

Chromatography and electrophoresis

Another method to fractionate a sample according to molecular size is *chromatography*. A particularly useful technique is gel filtration or gel exclusion chromatography. A solute in some kind of solvent is allowed to run through a column containing a loosely packed resin, such as Sephadex (cross-linked poly-dextrans) or Biogel (cross-linked polyacrylamide). Since the resin is porous, it acts as a molecular sieve, causing some of the solute to enter the pores and other solutes to go through. If a thin band of a solution containing one or more solutes is run onto the top of a column of resin and solvent is then allowed to flow through the column at a rate slow enough to allow solutes to equilibrate fully between the pore volume and the external volume the larger size solute will travel more slowly to the bottom than the smaller ones. The elution volume, which is the volume of solvent that has flown through before the first species emerges, is a measure of the partition coefficient σ, defined as the ratio of the solute concentration in the pores to the external volume. From the value of σ estimates of the size and the molecular weight of the solute can be made.

The movement of electrically charged macromolecules in a fluid as a result of an applied electric field is the basis of *electrophoresis*. If a macromolecule bears a net electric charge q, an applied electric field E results in a force $F = qE$ on the molecule. This force causes an acceleration of the molecule in a fluid until its velocity reaches a steady-state value v. At this velocity frictional forces are equal and opposite to the electric force F. Using Eq. 3.48 we have

$$v = qE/f \tag{3.64}$$

We can define the *mobility* u of the macromolecule as the velocity per unit electric field, *i.e.* $u = v/E$. Thus, using Eq. 3.49, the mobility of a spherical macromolecule with radius a and a net charge ze (e being the charge of an electron) would be

$$u = ze/6\pi\eta a \,. \tag{3.65}$$

Unfortunately, the theory of electrophoretic movement of biological macromolecules is much more complicated than the description just given would suggest. It appears that quantitative structural data cannot be expected. However, the technique can be used, and is very practical, for qualitative analysis and separation of proteins and nucleic acids.

A technique that recently has become very popular is electrophoresis of proteins in sodium dodecyl sulfate (SDS). This technique allows good estimates of molecular weight. SDS is an effective protein denaturant. It binds to all proteins in a mixture qualitatively in the same way; the same amount of SDS per amino acid is bound. The structures of the SDS-protein complexes are surprisingly similar. Hydrodynamically, the complexes appear to be prolate ellipsoids or rods with a constant diameter (about 18Å) and a length that is proportional to the molecular weight. Moreover, each SDS molecule has only one negative charge. The resulting charge density due to all SDS in the SDS-protein complex therefore overwhelms the variety of charge densities on the protein.

Many careful quantitative measurements have shown that the mobility is a linear function of the logarithm of the molecular weight. The equation

$$u = b - a \log M \qquad (3.66)$$

in which M is the molecular weight and a and b are constants depending on the gel concentration and the properties of the gel pore-size distribution, describes extremely well the experimental results (see Figure 3.41). For a particular gel (generally, polyacrylamide gels are used as support medium) at a given concentration a column can be calibrated by running proteins with known molecular weight through the column. The molecular weights then can be read from a calibration curve like the one given in Fig. 3.41.

3.7 X-ray crystal structure analysis

X-ray diffraction
The techniques discussed in the previous section suffice for a description of

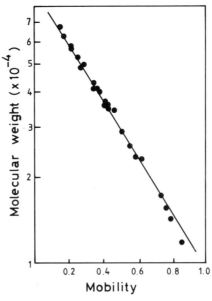

Fig. 3.41. Molecular weight as a function of the relative mobility of various proteins in 10% acrylamide gels. (From K. Weber and M. Osborn, The Proteins, vol. 1, (1975) Academic Press, New York.)

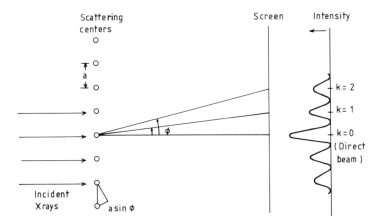

Fig. 3.42. Scattering of X rays from a single row of atoms. The intensity maxima occur as a result of constructive interference in the direction given by $a \sin \phi = k\lambda$, where a is the repeating distance between the scattering atoms.

gross size and shape of biological macromolecules. A detailed description of the macromolecular structure requires a different technique. X-ray crystal structure analysis is such a technique. X-rays are electromagnetic waves. If electromagnetic waves (light, for example) impinge on a regular array of scattering units (the lines of a grating, for example) diffraction occurs, and a diffraction pattern can be observed, if the distance between the scattering units is in the same order of magnitude as the wavelength of the waves. Thus, diffraction of X-rays, having wavelengths much shorter than light, can occur when the spacing of the scattering units is small enough. Since a crystal is a regular array of a unit structure (the unit cell), diffraction of X-rays by crystals occurs when the wavelength of the X-rays is of the same order of magnitude as the distances between the atoms in the crystal. In 1912 the group of Max von Laue in Germany showed that this indeed was the case. Thus, if, as in Figure 3.42, an X-ray beam is incident on a column of scattering units (atoms for example) in a crystal with a repeat distance a, a diffraction pattern results with maxima in the direction ϕ given by

$$a \sin \phi = k\lambda . \tag{3.67}$$

Crystal structure

A crystal is a three-dimensional successive translation of a basic unit structure, the *unit cell*, along three non-coplanar axes *a, b,* and *c.* It is a geometrical fact that three-dimensional space can be filled only by a mosaic of seven fundamental types of unit cell, each defining a crystal system. These seven types of unit cell can be grouped into fourteen so-called *Bravais lattices*, each being a unit cell with or without lattice points centered inside the unit cell (body-centered) or centered on two opposite, or all planes of the unit cell (face-centered). Figure 3.43 shows the fourteen Bravais lattices with the

95

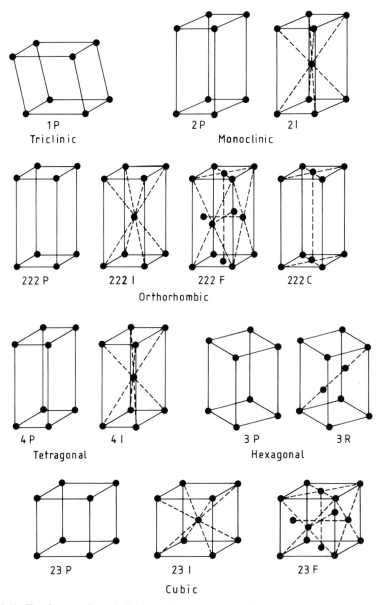

Fig. 3.43. The fourteen Bravais lattices. The seven types of unit cell are triclinic, monoclinic, orthorombic, tetragonal, hexagonal (primitive), hexagonal (rombohedral), and cubic. P is primitive, I is body-centered, F is face-centered, C is end-centered and R is rombohedral.

indications of the seven fundamental crystal types.

In order to define the faces of the crystal, or different sets of planes in the crystal, a set of indices has been developed. These indices, known als *Miller indices*, are defined as the reciprocals of the fractional intercepts of the planes with the *a*, *b*, and *c* axes of the unit cell. In Fig. 3.44 some planes in a cubic crystal are shown, illustrating the Miller indices. Planes which are

96

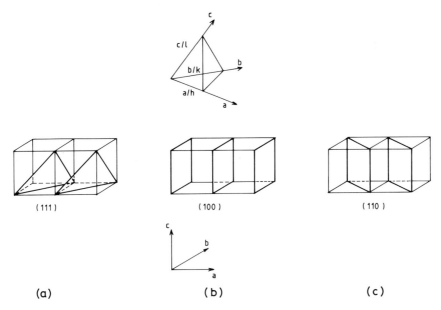

Fig. 3.44. Three examples of sets of planes given by Miller indices.

parallel to one or two of the axes intercept at infinity. The Miller index then is 0 (Figs. 3.44b and 3.44c).

The Bragg relation

In a three-dimensional array of scattering units, such as a crystal, the resulting pattern of maxima is too complicated to be analyzed in terms of a three-dimensional extension of the Von Laue theory. W.H. Bragg and his son W.L. Bragg developed a method which, essentially, reduces the three-dimensional problem to a two-dimensional one. They showed that the scattered X-rays could be seen as being reflected from planes in the crystal made up of the individual scattering units.

In Fig. 3.45 such planes, which can be indicated by the Miller indices, are represented by lines connecting rows of scattering units. Incoming X-rays, making an angle θ with these planes, are reflected at an equal angle θ. If the distance between two adjacent planes is d, constructive interference occurs when the difference in optical path length is an integer multiple of the wavelength. Hence, from Fig. 3.45, it follows that maxima are observed in the direction θ given by

$$2d \sin \theta = n\lambda \tag{3.68}$$

the so-called *Bragg relation*. The essential difference between Eqs. 3.67 and 3.68 is that in 3.68 the quantities d, θ, and λ are fixed in a resting system with monochromatic X-rays, whereas according to 3.67 a maximum will occur at a given scattering angle ϕ at any incident angle and at any value of a and λ, provided that the latter two values are compatible. According to 3.68

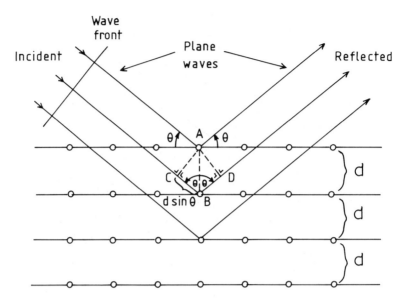

Fig. 3.45. Reflection of X rays from planes formed by scattering atoms. Constructive interference occurs in the direction given by an incident angle (which equals the reflection angle), satisfying the Bragg relation $n\lambda = 2d \sin \theta$, in which d is the distance between the reflecting planes.

however, a maximum occurs only when the incoming X-rays make an angle θ with a set of planes, satisfying the Bragg relation. Thus, although planes can be constructed in the crystal in a variety of directions, at a fixed direction of incidence no maxima will, in general, occur.

Diffraction patterns

X-ray diffraction patterns from crystals can be obtained in one of the following ways.

(a) By using nonmonochromatic X-rays the value of λ in Eq. 3.68 is made variable. The resulting diffraction pattern is known as a Laue pattern. Since the values of λ are not defined, the information obtainable from a Laue pattern is limited.

(b) One may use monochromatic X-rays (obtained by selecting a line of the characteristic radiation, such as the K_α line of Cu, with a suitable absorbing filter, such as Ni) on a mass of finely divided crystals with random orientations (a powder). In such a random-orientation arrangement some of the tiny crystals have orientations satisfying the Bragg relation 3.68. Since the reflected beam then makes an angle 2θ with the incident beam but the angles 2θ are oriented randomly around the incident beam, the maximum for each set of reflecting planes outline a cone. A photographic plate perpendicular with the direction of the incident beam will, therefore, show the maxima as concentric circles. These so-called *Debye-Scherrer rings* give direct information about the reflecting plane spacings.

(c) By rotating a single crystal in a fixed monochromatic X-ray beam the

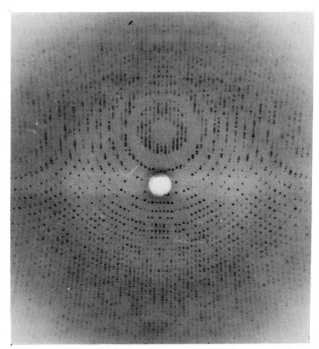

Fig. 3.46. Diffraction pattern taken from a crystal of a protein complex from a photosynthetic bacterium (from H. Michel (1982), J. Mol. Biol. **158**, 567–572; courtesy of Dr. H. Michel).

angle θ can be made variable. θ is measurable from the distance between the maxima showing on a cylindrical photographic film around the crystal. If the film is moved back and forth with a period synchronized with the rotation of the crystal, the positions of the spots on the film immediately indicate the orientations of the crystal. This technique is widely used for precise structure determinations. Since the value of λ is fixed, the plane distances d of the different sets of reflecting planes can be determined and thus, information can be obtained about the crystal structure. Also, the intensity of the spots, as will be seen later, contains important information about the fine structure of the crystal. A diffraction pattern taken from a crystal of a protein complex from a photosynthetic bacterium is shown in Fig. 3.46.

When a crystal is rotated in a beam of monochromatic X-rays about an axis parallel to one of its unit cell sides, the reflected radiation, according to Eq. 3.67, must outline a cone $\sin \phi$ = constant. These cones cut lines on a cylindrical film around the crystal. The maxima thus produce spots on the film which lie on these so-called layer lines. The vertical deflection of the layer lines is given by

$$\sin \theta_k = k\lambda/b \qquad (3.69)$$

in which b is the length of the b-axis of the unit cell (see Fig. 3.47). Thus, measurements of the layer line distance after rotation of the crystal about an axis parallel to the unit cell axes (a, b, or c) will yield the dimensions

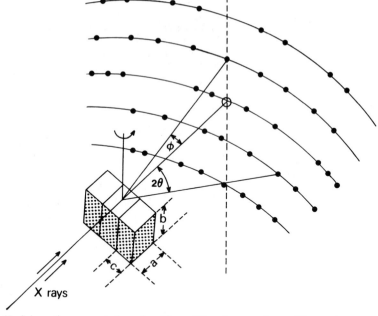

Fig. 3.47. Schematic representation of an X-ray diffraction experiment. The maxima are situated on layer lines. The vertical deflection of the layer lines is given by ϕ and the various diffraction maxima have deflections 2θ.

of these unit cell axes. The spacings between the spots on the film are determined, of course, by the Bragg relation 3.68 and indicate the reflecting plane distances *d*.

Intensity of diffraction maxima

As an example of the way in which the maxima can be described, consider a structure with scattering elements in all points of the simple cubic lattice illustrated in Fig. 3.48a. Monochromatic X-rays scattered from one (100) plane will be exactly in phase with those from successive (100) planes when the angle of incidence satisfies the Bragg relation. This is observed because all (100) planes contain all the scattering elements of the structure. The same is true for the (110) and the (111) planes. The lattice illustrated in Fig. 3.48b is one of a body-centered structure. In this case the (100) planes do not contain all the scattering elements of the structure. Hence, while X-rays coming in at the Bragg angle and reflected at the (100) planes reinforce each other, X-rays reflected from the interleaved planes (which are exactly out of phase with the others) will reduce the intensity. If the scattering elements all have the same scattering power the resultant intensity will be reduced to zero and no first-order (100) maximum will appear. If, however, the scattering elements are different, the first-order (100) maximum will still appear but will be reduced in intensity. The (110) planes of the body-centered structure do contain all the scattering elements and, therefore, a strong first-order (110) maximum

100

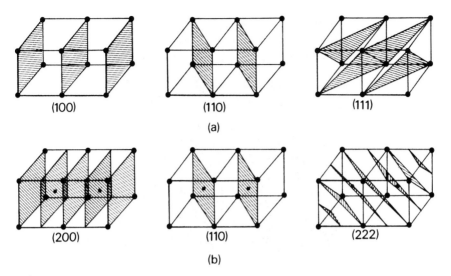

(100)   (110)   (111)

(a)

(200)   (110)   (222)

(b)

Fig. 3.48. Reflecting planes in (*a*) a simple cubic lattice and (*b*) a body-centred cubic lattice.

will show up; also, the second-order reflection at the (100) planes will show a strong maximum. This is the exact equivalent of the first-order reflection at the (200) planes which again contain all scattering elements of the structure.

Diffraction by helical structures

The application of diffraction techniques for a helical structure is shown in Fig. 3.49. In the figure the helix has a pitch of a and an integral number (six) of scattering elements per pitch. The first-order (100) reflection will be zero (when each of the scattering elements has equal scattering power) since each of the interleaving planes containing scattering elements will cause a reflection which is 1/6 out of phase. The same is true for the second-, third-,

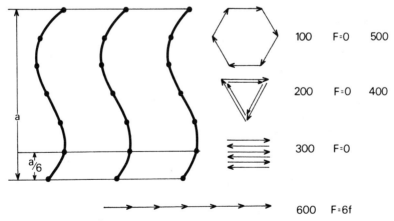

100   F=0   500

200   F=0   400

300   F=0

600   F=6f

Fig. 3.49. A helical structure with pitch a and six scattering elements per pitch. The vector diagrams show how the maxima of a diffraction pattern are formed.

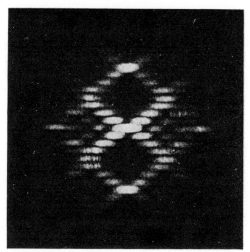

Fig. 3.50. An optical simulation of a diffraction pattern from a helical structure, such as the one showed in Fig. 3.49 (from A.R. Stokes, in Progress in Biophysics, 1955, Pergamon Press, London and New York).

fourth-, and fifth-order (100) reflections (of course equivalent to the first-order (200), (300), (400), and (500) reflections, respectively. The first-order (600) reflections will then show a strong maximum. This is illustrated quite clearly when the vector description is used; the scattered radiation has amplitude and phase and, therefore, can be represented by vectors. In the case of the helix of Fig. 3.49 the vector from each scattering element has a $2\pi/6$ phase difference with the preceding one. The resulting diffraction pattern will be something like the one shown in Fig. 3.50.

Phases

As can be seen from the preceding examples, the intensity of a diffraction maximum is determined not only by the amplitude of the scattered X-rays but also by their phases. In an actual crystal the lattice points are usually occupied not by a single scattering element but by a group (actually, an electron density distribution). As an example consider (see Fig. 3.51a) a structure formed by replacing each point in a lattice by two atoms, a black one and a white one. If a set of reflecting planes is drawn through the black atoms another parallel, but slightly displaced, set can be drawn through the white atoms. When monochromatic X-rays are incident on the set of planes at the Bragg angle the reflections from all the black atoms will be in phase and the reflections from all the white atoms will be in phase. The radiation from the black atoms, however, is slightly out of phase with the radiation from the white ones, so that the resulting amplitude will be diminished (Fig. 3.51b).

To find a general expression for the phase difference, consider the two-dimensional cross section illustrated in Fig. 3.51c. The black atoms are placed in the corners of a unit cell with sides a and b, and the white atoms have displaced positions. Take the coordinates of the black atoms as (0,0) and

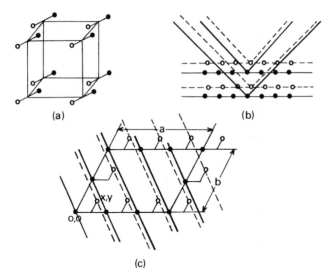

(a)

(b)

(c)

Fig. 3.51. (a) Simple cubic lattice with two scattering elements at each point; (b) Bragg reflection at a set of planes from the cubic lattice of (a); (c two-dimensional cross section of a simple lattice.

of the white atoms as (x,y). A set of planes hk is shown for which the Bragg condition is fulfilled. The spacings a/h along a, and b/k along b correspond to positions for which the phase difference is 2π rad, *i.e.* the scattering from these positions is exactly in phase. The phase difference between these planes and those going through the white atoms is proportional to the displacement of the white atoms. The phase difference φ_x for the displacement x in the a direction is given by $x/(a/h) = \varphi_x/2\pi$ or $\varphi_x = 2\pi h(x/a)$. The total phase difference for the displacement in both the a and the b directions becomes

$$\varphi_x + \varphi_y = 2\pi[h(x/a) + k(y/b)]$$

and by extension to the three-dimensional case,

$$\varphi = 2\pi[h(x/a) + k(y/b) + l(z/c)] .$$ (3.70)

Superposition of waves with different amplitudes and different phases is best accomplished by vector addition. If f_1 and f_2 are the amplitudes of the waves scattered by atoms 1 and 2, respectively, and φ_1 and φ_2 are their phases, the resulting amplitude will be $F = f_1 e^{i\varphi_1} + f_2 e^{i\varphi_2}$. For all the atoms in the unit cell this becomes

$$F = \sum_j f_j e^{i\varphi_j} .$$ (3.71)

When (3.71) is combined with (3.70) we obtain the resulting amplitude of the waves scattered from the (hkl) planes by all the atoms in the unit cell

$$F(hkl) = \sum_j f_j e^{2\pi i(hx/a + ky/b + lz/c)}$$ (3.72)

103

This quantity is called the *structure factor* of the crystal. It is dependent on the positions of the atoms and the atomic scattering factors f_j. The values of the atomic scattering factors are given by the scattering angle and the number and distribution of the electrons in the atom; it is the electrons in a crystal structure that scatter the X-rays.

Fourier analysis

A crystal examined by X-rays may be regarded as a periodic three-dimensional distribution of electron density, $\rho(xyz)$; such a density distribution can be expressed as a Fourier series. It can be shown that the Fourier coefficients are then the structure factors divided by the volume V of the unit cell. Thus,

$$\rho(xyz) = \frac{1}{V} \sum F(hkl)e^{-2\pi i\,(hx/a\,+\,ky/b\,+\,lz/c)}. \tag{3.73}$$

The summation is carried out over all values of h, k, and l, so that there is one term for each set of planes (hkl) and, hence, for each spot on the X-ray diffraction pattern.

The phase problem

Equation (3.73) summarizes the whole problem of structure determination. Since the crystal structure is simply the electron density distribution $\rho(xyz)$, positions of individual atoms are peaks in ρ with heights proportional to the atomic numbers (numbers of electrons). If $F(hkl)$ is known, the structure could be immediately plotted. However, we can only measure the intensities of the spots, which are proportional to $|F(hkl)|^2$; we can calculate the amplitudes but we have lost the phase differences in taking the X-ray scattering pattern. This inability to measure phase differences experimentally poses a serious problem, especially for biological macromolecules which produce an enormous number of diffraction spots (*cf.* Fig. 3.46). A number of techniques have been developed to help infer partial phase information. Often, chemical insights into the structure or other previously known information (such as, for instance the number and approximate locations of chromophores) must be included to assist in the structure determination.

A procedure to determine the structure of relatively small molecules (with molecular weights of less than a few thousands) starts with trying to estimate the locations of a few atoms. If such atoms are "heavy", that is if they have a relatively large atomic number, their locations can be found by making use of the so-called Patterson function. A Patterson function is a Fourier transform, not of the structure factor, like in Eq. 3.73, but of the intensity. This function is, in fact, a *vector map* of the structure. Unfortunately, there is no simple way of deciding which pair of atoms belongs to a given vector. However, since the peaks in the Patterson function are proportional to the product of the atomic numbers of every pair of atoms the vectors between the pairs of heavy atoms in the structure are the most dominant feature of the Patterson map. Thus, with a limited number of heavy atoms it is usually possible to find out enough about their locations to make a good estimate of the phase differences. Using these *calculated* phase differences together with the *measured* amplitudes, one can make a Fourier transfer, leading to an electron density distribution. This density distribution in turn can be used to calculate more accurate phase differences which then are used in a Fourier synthesis giving a new electron density distribution. This procedure, which is called a *Fourier refinement*, is repeated until the locations of all atoms are consistent with expectations based on known information about the structure. Sometimes an additional refinement is necessary by allowing the structure to vary somewhat and maximize the agreement between the computed structure and the observed data by a least-square method.

The procedure described above turns out not to be successful with large complex macromolecules, such as proteins or nucleic acids. These molecules generally do not contain conveniently located heavy atoms. Also, the complexity of their structures require other approaches. A successful approach is known as the *multiple isomorphous replacement technique*. This technique consists of preparing crystals which are identical to the crystals of the macromolecule (the parent crystal), except that one or more heavy atoms are introduced at specific loci. By comparing diffraction patterns of at least two heavy atom isomorphous derivate crystals with that of the parent crystal an estimate of the location of the heavy atoms can be made. Using these positions and the differences between the intensities of the diffraction spots of the parent crystal and the heavy atom derivates, it is possible to estimate the phase of the structure factors. These estimates can be made more precise by repeating the procedure or by refining methods similar to those indicated above for small molecules. Finally the phase differences can be used to compute an image of the structure.

In spite of the long and tedious work involved, the elucidation of the structures of a number of proteins has been successful. Pioneering successes have been the determination of the structure of myoglobin by J.C. Kendrew, of hemoglobin by M.F. Perutz, and of vitamin B_{12} by Dorothy Hodgkin. The method has been employed in a dramatically successful way by F.H. Crick and J.D. Watson and by M. Wilkins in the discovery of the double helix structure of DNA, as a result of which the molecular mechanism of heredity could be explained (see Chapter 4). The development of powerful computers and software, not only to do the calculations but also to plot the results in various forms that each emphasize particular aspects of the structure, has contributed substantially to the successes of the method.

A recent example is the elucidation, for the first time, of the detailed structure of a complicated membrane protein, a Reaction Center of a photosynthetic bacterium (Chapter 6). In Fig. 3.46 the diffraction pattern of a crystal prepared from this complex is shown. The protein consists of four subunits. One of them, a cytochrome subunit, has four heme chromophores and two others, the L and M subunits, share four chlorophylls, two pheophytins (chlorophyll without the central Mg atom) and some additional prosthetic groups. The composition of the protein complex was already known from spectroscopic and biochemical data. As usually is the case, it took time and hard work

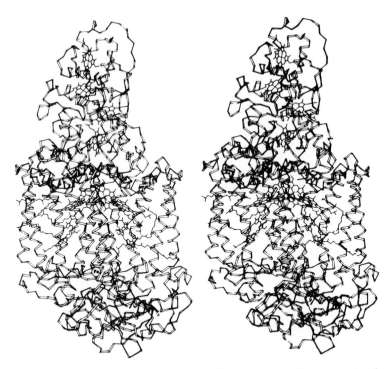

Fig. 3.52. A stereophotograph of the structure of the Reaction Center complex from the photosynthetic bacterium *Rhodopseudomonas viridis*. In order to view the photograph in stereo: hold the picture at 40 to 50 cm from your eyes and focus on a distant point *behind* the picture. From the three images, the middle one should have a stereoscopie appearance (from J. Deisenhofer *et al*, Nature **318**, (1985) 618–624); courtesy of Drs. H. Michel and J. Deisenhofer).

to get suitable crystals and to prepare the heavy atom derivates. The structure was finally determined with a resolution of 3 Å by analyzing between 20,000 and 50,000 spots each from 6 parent crystals and from a total of 28 crystals from five different heavy atom derivates. The structure, which is shown in Figure 3.52, confirms the previously known spectroscopic and biochemical data and adds substantially to our insight in the way of how membrane proteins function. The cylindrical central part of the complex (bottom half of the picture) contains the subunits L and M and is about 65Å long. The subunits both contain long helical regions (see Chapter 4) consisting mainly of hydrophobic amino acid residues. Most likely, therefore, this part spans the membrane. The major part of the other subunit, the H-subunit, is in contact with the outer flat surface of the L-M complex (directed to the outside of the vesicle which is enclosed by the membrane). The other flat surface of the L-M complex binds the cytochrome subunit (top half of the picture), consisting of more hydrophylic residues; it is in contact with the aqueous phase at the inside of the membrane bound vesicle. In Chapter 6, this fascinating protein complex is described in somewhat more detail in connection with its function in the photosynthesis process.

Bibliography

Cantor, C.R., and Schimmel, P.R. (1980). "Biophysical Chemistry part II: Techniques for the Study of Biological Structure and Function." W.H. Freeman and Cie., San Francisco.

Chen, S.H., and Yip, S. (1974) "Spectroscopy in Biology", Acad. Press, New York.

Deisenhofer, J., Epp, O., Miki, K., Huber, R., and Michel, H. (1984). X-ray Structure Analysis of a Membrane Protein Complex, *J. Mol. Biol.* **180**, 385–398.

Deisenhofer, J., Epp, O., Miki, K., Huber, R., and Michel, H. (1985). Structure of the Protein Sunbunits in the Photosynthetic Reaction Centre of *Rhodopseudomonas viridis* at 3 Å Resolution, *Nature* **318**, 618–624.

Ehrenstein, G., and Lecar, H. (Eds.). (1982). "Biophysics", Vol 20 of "Methods of Experimental Physics" (R. Celotta and J. Levine, eds. in chief), Acad. Press, New York.

McMillan, W.S., and Mayer, J.E. (1945). The Statistical Dynamics of Multicompartment Systems, *J. Chem. Phys.* **13**, 276–305.

4. Structure and function of proteins and nucleic acids

4.1 The structure of proteins

Structure and function
Shortly after the discovery of X-ray diffraction by crystals the technique was used to investigate biological macromolecules, in particular proteins which could be crystallized and nucleic acids. The overriding importance of such studies is that they show the close relationship between the structure and the function of these macromolecules. Proteins have a greatest diversity of functions in an organism. These functions are performed for the most part by selective binding to molecules. The selectivity is assured by numerous weak interactions working at close range between substrate (or *ligand* as it is called frequently) and macromolecule, so that binding only can be sufficiently tight if there is close fitting of the ligand to the protein.

   If binding links identical molecules, so that many copies of the same protein aggregate, large-scale structures are formed such as fibers or tubules. Often, however, the binding involves molecules different from the binding protein. Enzymes are the most obvious examples. But also other protein functions involve selective binding, such as binding between antigen and antibody, binding of regulating proteins to parts of DNA, binding of ligands to receptor proteins for endocytosis or recognition for the purpose distinguishing self from nonself, and so on. Virtually all the activities of proteins can be understood in terms of such selective binding. Knowledge of the structure, therefore, is imperative to understand its function.

Astbury structures
Around 1930 Astbury tried to interpret the diffraction pattern obtained from crystallized protein and nucleic acid fibers. He showed that protein fibers give rise to two types of patterns which he called α and β patterns. He recognized that the α pattern was due to a more folded and dense structure (Fig. 4.1a) while the β pattern was from a more stretched structure (Fig. 4.1b). The structures in his models are held together by hydrogen bonds which are indicated in Figure 4.1 by dotted lines.

The α-helix and β-pleated sheet
Although the importance of Astbury's models for the understanding of protein structure cannot be denied, they did not give a completely satisfactory explanation of the diffraction patterns. The helical structure proposed by

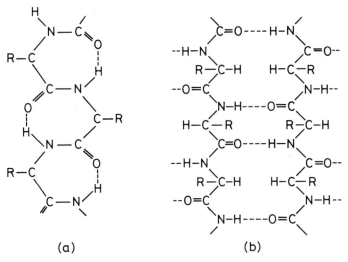

Fig. 4.1. Protein structures according to Astbury; (*a*) α structure and (*b*) β structure.

Pauling and Corey in 1951 did fit the diffraction patterns of many fibrous proteins much better. Astbury and others have also tried helical models but they never thought of fitting a noninteger number of amino acid residues into one turn. Taking into account the known bond distances and bond angles, Pauling and Corey showed that a helical structure with 3.6 amino acid residues per turn, a diameter of about 6.8 Å, and a distance between turns of about 5.4 Å would be a definite possibility. The helical structure is stabilized by hydrogen bonds between amino acids four units apart. These hydrogen bonds are between the amino group of one peptide unit to the carboxyl group of another and do not involve the residues. The stability of the α helix, therefore, is not depending on the identity of the residues which form side chains pointing out from the helix. Thus, the same α-helical structure can accommodate almost any flexible side chain as long as they are L-amino acids and not too long to interfere sterically with the structure. Proline is an exception, however. Due to its peculiar structure (see Table 2.1), this amino acid never occurs in an α helix. It causes more or less sharp bends in the peptide chain.

This α-helical structure fits the diffraction patterns observed for synthetic polypeptides very well and there were also many aspects pointing to the α helix in the diffraction patterns of many naturally occurring fibers of the α-type. The model, which is illustrated in Fig. 4.2a, turned out to be remarkably accurate, considering the fact that it was proposed six years before an α helix was actually seen at molecular resolution for the first time in the crystal structure of myoglobin. Since then it is confirmed in many structures of biological macromolecules.

Pauling and Corey also made pleated sheet models in line with Astbury's β structure. They did not restrict the peptide bonds to one plane (Fig. 4.2b). Both structures, the α helix and the pleated sheets, retain the idea of an α structure being folded and a β structure being stretched out.

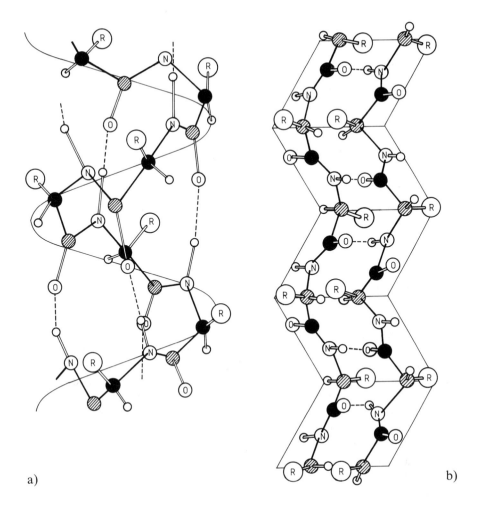

a)

b)

Fig. 4.2. (*a*) A drawing of the helix polypeptide chain according to Pauling and Corey. The black circles represent the C_α atoms, the shaded circles the other C atoms; the small circles without a letter represent hydrogen atoms; the dotted lines designate hydrogen bonds. R are the amino acid residues. The helix has a radius of 3.4 Å and an interturn distance of 5.4 Å. There are 3.7 residues per turn. (*b*) Two antipatallel β strands in a pleated sheet configuration.

Generally proteins contain both α helix and β sheet. Some are composed mainly of α helices, others are predominantly β sheet. A typical protein can have an interior of bundles of β strands running diametrically back and forth while its surface is covered with α helices.

Tertiary structure

In numerous proteins one can also see structural features consisting of two β strands connected by a stretch of α helix. These three pieces fit nicely if they are arranged at particular angles. This kind of structure is an example

111

of what is called a *domain*. When two such domains are adjacent, the crevice they form often is a binding site. The α–β–α domain, therefore, is of particular importance.

The tertiary structure of a protein can be seen as a geometric arrangement of domains. Such a tertiary structure is consistently formed by a particular amino acid sequence but this is not true in general. Artificially constructed random sequences of nonphysiologival amino acids are generally loose and flexible coils that continually shift from one structure to another. However, the subset formed by the twenty L–amino acids ocurring in living organisms always seems to result in unique and stable structures. If sufficient mathematical tools were available and if all factors, such as hydrogen bonding and hydrophobic interaction were well enough understood it should be possible to predict the complete three–dimensional structure of a protein from its amino acid sequence.

Quaternary structure
In many proteins two or more units of tertiary structure (subunits) are bound together by noncovalent interaction into a larger quaternary structure. In the simplest cases two identical subunits are bound in one protein, an example being liver alcohol dehydrogenase. One of the most studied proteins, mammalian hemoglobin, consists of four subunits which form two identical pairs (designated as $\alpha_2\beta_2$). The function of hemoglobin is to transport oxygen in the bloodstream. It can therefore be in the oxygenated state (oxyhemoglobin) or in the deoxygenated state (deoxyhemoglobin). Figure 4.3 gives a schematic view of horse deoxyhemoglobin. Two α subunits and two β subunits form two identical pairs of subunits. Each subunit contains a heme group, a closed so-called tetrapyrole ring containing an iron atom in the Fe^{2+} state. In deoxyhemoglobin each heme iron is coordinated with five ligands; four of these are in the tetrapyrole ring itself whereas the fifth comes from a histidine of the subunit polypeptide chain. Upon oxygenation, the sixth ligand position on the iron is coordinated with oxygen, resulting in a change of the conformation of the molecule. By virtue of this conformational change the protein functions as an *allosteric enzyme* (see next section).

Many more complicated quaternary structures of proteins are known with more than one of several kinds of subunits. The enzyme RNA polymerase from the colon bacterium *E. coli*, for example, is composed of five subunits of three different kinds ($\alpha_2\beta\beta'\sigma$). A very large complex is the pyruvate dehydrogenase complex, also from *E. coli*; it is composed of 24 different kinds of subunits and has a molecular weight of 5,000,000.

4.2 Enzymes

Catalizing function of enzymes
Enzymes are proteins in which the relation between structure and function is clearly visible. These proteins speed up the rate of a particular reaction

112

or group of reactions, but do not change the equilibrium conditions. They enter the reaction but are in the same state after the reaction as they were before. The oxidation of glucose, for example, can be represented by the equilibrium equation

$$C_6H_{12}O_6 + 6O_2 \rightleftarrows CO_2 + 6H_2O .$$

The equilibrium strongly favors the forming of CO_2 and H_2O, because the potential energy of the carbon dioxide and the water together is much lower than that of the sugar and the oxygen together. However, when the glucose and oxygen are brought together, for instance in a solution, no reaction will take place because the reaction rate is too slow. This is the case since going from the left–hand side of the equation to the right–hand side the system has to go through an intermediate state with a potential energy much higher than even the initial state (as illustrated in Fig. 4.4). The difference between the initial potential energy and that of the intermediate state is called the activation energy. At room temperature or at slightly elevated temperatures, such as that inside a human body, only very few molecules have enough energy to overcome this barrier. Heating often results in a speeding up of the reaction by increasing the initial energy of the molecules (one can "burn" sugar). Enzymes have the effect of lowering the activation energy. They do this by forming noncovalent complexes with the substrate.

There are many kinds of enzymes which are grouped according to the kind of reaction they catalyze. The *hydrolases* catalyze hydrolysis reactions (the splitting of a molecule, adding H to one part and OH to the other), the *transferases* catalyze the transfer of a piece of one molecule to another, the *isomerases* accelerate isomerizations, the *carboxylases* remove CO_2, and the

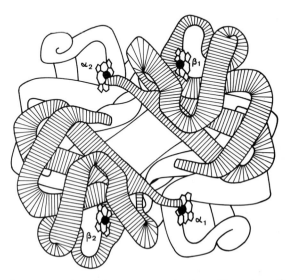

Fig. 4.3. A schematic representation of the quaternary structure of horse hemoglobin.

respiratory and photosynthetic enzymes catalyze the oxidation–reduction reactions of respiratory and photosynthetic processes.

Coenzymes, cofactoes and prosthetic groups

Many enzymes consist of a protein molecule as well as a smaller molecule. In this case the large protein part is called *apoprotein* (or apoenzyme). The smaller part, dependent upon its nature, is called *coenzyme* (a large organic molecular group which separates easily from the enzyme), a *cofactor* (a loosely bound small inorganic ion), or a *prosthetic group* (a small nonprotein part of the enzyme which is attached so firmly that it cannot be removed easily without irreversibly changing the enzyme). An example of a prosthetic group is the heme group, containing iron, in the oxidative enzymes as hemoglobin, myoglobin, and the cytochromes.

The specifity of enzymatic reactions is contingent upon the formation of a complex of enzyme and *substrate* (the substance or group of substances that undergo biochemical conversion). Since the complex is formed by noncovalent interactions, there must be a relative large number of points at which this interaction can take place in order to ensure a certain amount of stability for the complex. Moreover, the noncovalent interaction must be at close range, in other words there must be many places at which atoms of the substrate can come very close to atoms of the enzyme. Thus, a structural correspondence must exist between the substrate and the enzyme. It is this very fact that lies at the basis of recognition processes, not only when enzymes are involved, but also in reactions involving receptor proteins, such as endocytosis and immunological processes (see below). The idea is illustrated in Fig. 4.5. Although the drawn shapes should not be taken literally, the key-keyhole concept explains the enzyme reaction kinetics excellently. According

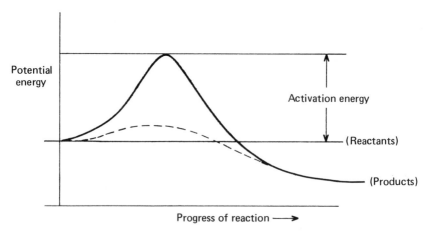

Fig. 4.4. The activation energy barrier. Although the potential energy of the products of a reaction is lower than that of the reactants, the reaction cannot proceeds at measurable speed because it has to go through a state of high potential energy. When the reaction is catalyzed (by an enzyme, for example) the potential energy barrier is substantially lowered (dashed line).

114

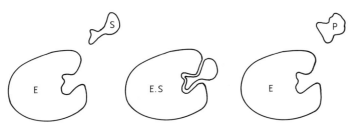

Fig. 4.5. A representation of enzyme action. The substrate S can form a complex E.S with the enzyme E because its configuration "fits" at the active site of the enzyme. The substrate can then react (for instance by hydrolysis), thus forming the product and free enzyme.

to this concept the action of competitive inhibitors (Fig. 4.6) can also be analyzed.

Enzyme kinetics
The fact that the action of enzymes is based on the formation of a substrate–enzyme complex was already established by Michaelis and Menten, who concluded this from the reaction kinetics of enzymatic reactions. For a simple enzymatic reaction involving only one substrate and yielding a single product a plot of the reaction rate against the substrate concentration is a curve such as those illustrated in Fig. 4.7. The rate starts to increase linearly but levels off to a plateau. The height of the plateau depends on the concentration of the enzyme. One can derive such a curve by considering the following reaction

$$E + S \underset{k_{-1}}{\overset{k_1}{\rightleftarrows}} E.S. \underset{k_{-2}}{\overset{k_2}{\rightleftarrows}} P + E \tag{4.1}$$

in which is E is the enzyme, S the substrate, and P the reaction products. The equilibrium of the second reaction usually is far to the right, hence the back reaction rate constant k_{-2} is negligible. Then, after a while, if $[S] > [E]$

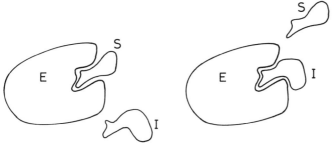

Fig. 4.6. Competitive enzyme reaction inhibition. The inhibitor I has a configuration which is similar to that of the substrate. The inhibitor can bind to the enzyme, thereby blocking the formation of the enzyme-substrate complex.

115

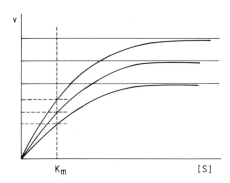

Fig. 4.7. A plot of the reaction rate v (= $d[P]/dt$), as a function of the concentration of substrate [S] at different enzyme concentrations. At half of the saturation rate the substrate concentration is equal to the Michaelis constant K_m.

(which is usually the case), the reaction rate

$$v = d[P]/dt = k_2[\text{E.S.}] \qquad (4.2)$$

becomes constant, so

$$d[\text{E.S}]/dt = 0 = k_1[\text{E}][\text{S}] - k_{-1}[\text{E.S}] - k_2[\text{E.S}] . \qquad (4.3)$$

Since the concentration of free enzyme [E] is equal to the maximum concentration $[\text{E}]_0$ minus the concentration of the complex,

$$[\text{E}] = [\text{E}]_0 - [\text{E.S}] \qquad (4.4)$$

Eq. (4.3) becomes

$$k_1\{[\text{E}]_0 - [\text{E.S}]\}[\text{S}] = (k_{-1} + k_2)[\text{E.S}] . \qquad (4.5)$$

We now define

$$K_m = \frac{k_{-1} + k_2}{k_1} . \qquad (4.6)$$

From (4.5) and (4.6)

$$K_m = \frac{[\text{E}]_0 - [\text{E.S}]}{[\text{E.S}]}[\text{S}] \qquad (4.7)$$

and

$$[\text{E.S}] = \frac{[\text{E}]_0[\text{S}]}{K_m + [\text{S}]} . \qquad (4.8)$$

116

The constant K_m is called the *Michaelis constant* and describes, in a relatively simple way, the enzyme reaction kinetics. If

$$[S] \gg K_m \tag{4.9}$$

Eq. (4.8) becomes

$$[E.S.] = [E]_0 \text{ and } v = k_2[E]_0 = v_m \tag{4.10}$$

in which v_m is the maximum rate. Thus at high substrate concentration the rate is limited by the maximum amount of enzyme. All of the enzyme is bound in the complex and there is no free enzyme left. The rate is independent of the substrate concentration. If, however,

$$[E] \ll K_m \tag{4.11}$$

Eq. (4.8) becomes

$$[E.S] = \frac{[E]_0[S]}{K_m} \tag{4.12}$$

and the rate is

$$v = \frac{k_2[E]_0[S]}{K_m} . \tag{4.13}$$

The reaction is limited by the rate at which the enzyme and the substrate can form the complex.

At high substrate concentration the reaction is of the zeroth order (rate is constant). This is the plateau in the curves of Fig. 4.7. Since the maximum rate $v_m = k_2[E]_0$,

$$v = k_2 \frac{[E]_0[S]}{K_m + [S]} = \frac{v_m}{K_m/[S] + 1} \tag{4.14}$$

which shows that the reaction rate is at half maximum when $[S] = K_m$.

The parameters K_m and v_m are most conveniently obtained by rearranging Eq. 4.14. This can be done in two ways:

$$1/v = (K_m/v_m)(1/[S] + 1/v_m) \tag{4.15}$$

and

$$v/[S] = -v/K_m + v_m/K_m . \tag{4.16}$$

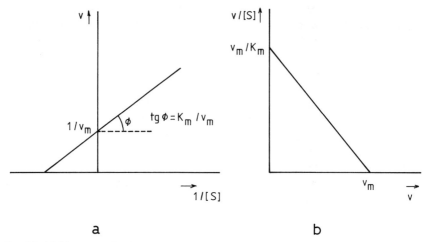

Fig. 4.8. (a) Lineweaver-Burk plot; (*b*) Eady-Hofstee plot.

When Eq. 4.15 is used $1/v$ is plotted versus $1/[S]$. This plot is called a *Lineweaver–Burk* plot (Fig. 4.8a). The slope of this plot gives K_m/v; the intercept with the $1/v$ axis gives $1/v_m$. In using Eq. 4.16 $v/[S]$ is plotted versus v which gives and *Eady–Hofstee* plot (Fig. 4.8b). The intercepts of this plot with the vertical and horizontal axes give v_m/K_m and v_m. Because each method of plotting uses the data points differently, it is advisable to determine the parameters by using both plots and compare the results.

Allosteric effect
Another structure-related functional effect is the allosteric effect. If, for instance, the rate of the binding of oxygen by hemoglobin is plotted as a function of the partial pressure of the oxygen, a sigmoidal (S-shaped) curve is obtained (Fig. 4.9). This means that the affinity of hemoglobin for oxygen

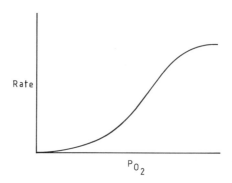

Fig. 4.9. A plot of the oxygenation of hemoglobin as a function of the partial pressure of oxygen. The sigmoid shape of the curve shows that the reaction goes faster with the more oxygen bound. This is due to the allosteric action of the enzyme.

118

is increased as more oxygen is bound by the molecule. Heme-heme interaction cannot explain this effect because of the relatively large distance between the heme groups in the subunits. The interaction must be indirect, that is through the binding sites of the subunits. Conformational changes in the protein structure which alter the affinity for specific substrates result from this allosteric effect.

Allosteric effects now have been found in many enzymes. The effect is very important for the regulation of enzymatic processes; sometimes the effect is stimulating (as with the case of hemoglobin), sometimes the effect is inhibitory; sometimes the products, or their degradation products, excert the allosteric effect, sometimes the precursors exert the effect. It is through such processes, and often through combinations of them, that the enzymatic processes in a living cell are very effectively regulated.

4.3 Recognition proteins

Specific binding
In many life functions recognition proteins are of vital importance. In a certain sense the enzymes, which we discussed in the previous section, are also recognition proteins; the substrates must be recognized by the enzyme before they are processed. We have briefly met other recognition proteins when we discussed receptor mediated endocytosis (Chapter 2). In that case specific ligands are recognized by proteins on the membrane surface; by binding these recognized ligands, the proteins cause the membrane to draw the complex inside. In many instances this is the way in which specific proteins, synthesized elsewhere in the organism, are brought into cells where they are needed. An example is the formation of egg yolk in bird eggs. Specific proteins which are synthesized in the liver of the female bird are transported by the blood to the ovary where they bind to receptor proteins on the membrane of the developing oocyte (the precursor cell of an egg). The membrane invaginates and forms vesicles inside the cell. These vesicles then fuse to form the egg yolk. Another example is the way in which iron is brought into cells by a carrier protein called *transferrin*. Two iron ions bound to a transferrin molecule are imported into a cell by endocytosis triggered by the binding of the transferrin to specific receptor proteins on the cell surface.

The immune system
In no case is the importance of recognition proteins so obviously demonstrated as in the immune system of the higher organisms. The cells and the molecules of this system form a defensive network in which recognition, not only of foreign invading cells and substances but also native cells plays a vital role.

The important cells of this system are the *lymphocytes*, or white blood cells. In mammals they come in two classes. In one of these classes are the *B* cells which mature from stem cells in the bone marrow. These cells produce the *antibodies* or *immunoglobulins* which specifically bind to foreign substances, called *antigens*. An antibody clearly is a recognition protein. The other class contains the *T* cells which also originate from bone marrow stem cells but they mature in the thymus gland where they undergo further differentiation. These cells produce recognition proteins, called *T cell receptors* which enable them to recognize only those cells that bear certain marker proteins. By this means the *T* cells are given the possibility both to act against viral infection and to regulate other components of the immune system. Also receptor and marker proteins are clearly recognition proteins. All the components of the system cooperate in a network of positive and negative feedback to fend off infection from foreign invaders. In this book, an extensive discussion of this rather complicated network cannot be given; we would like to refer the interested reader to the review articles on this subject (see Bibliography). Here we discuss only a few aspects of the system in order to demonstrate the relation between the structure and function of the recognition proteins of the system.

Antibodies are produced by *B* cells and displayed on the membrane surface. When the cell meets an antigen which is specific for its antibody, the antigen is bound and drawn into the cell where it is processed. A smaller processed piece of the antigen is then displayed at the cell surface in combination with a membrane protein that marks the *B* cell as belonging to that particular organism. There are two classes of such marker proteins. They are called *MHC* proteins, *MHC* standing for Major Histocompatible Complex.* Class I *MHC* proteins are found on the surface membranes of all (nucleated) cells of the organism, class II *MHC* only on the surface membranes of lymphocytic cells. The complex of the antigen part with class II *MHC* protein on the membrane surface of a *B* cell is recognized by a receptor protein on the surface membrane of a *T* cell. Binding of the *T* cell to the antigen presenting *B* cell triggers the *B* cell to proliferate in two directions. One line of progeny produces new cells, the so-called *plasma* cells, that make many copies of the original antibody, this time not bound to the membrane but released as free antibody that can bind to free antigen, thus marking it for destruction. The other line is a clone of *B* cells that remain circulating in the blood, thus forming some sort of a memory for the system to deal more efficiently with the next invasion of the same antigen.

The *T* cells that function this way are the so-called *helper T* cells. There are also *T* cells that can kill cells. These are the so-called *cytotoxic T cells.*** They are effective in killing cells infected by a virus, thus preventing the repoduction of the virus. The cytotoxic *T* cell also has a receptor protein on its surface. That receptor protein can recognize an antigen produced by

---

* The correct name, in fact is *MHC-encoded protein*. The *MHC* is a collection of closely linked pieces of DNA that encode for the protein (see section 4.5).

** In addition to helper and cytotoxic *T* cells there are also suppressor *T* cells. These cells are instrumental for the "fine tuning" of the system.

a virus only if it is bound to a class I *MHC* protein on the surface membrane of the infected cell. The cytotoxic *T* cell then can bind to the infected cell and kill it. The antigen-*MHC* combination that is required for the killing prevents the killing of the organism's own cells. On the onther hand, A "non-self" *MHC* protein, on the surface of a foreign cell for instance, is seen by the *T* cell receptor as an antigen combined with a "self" *MHC* protein; the foreign cell, therefore, is attacked and killed by the cytotoxic T cell. Hence the problem of rejection in organ transplantation.

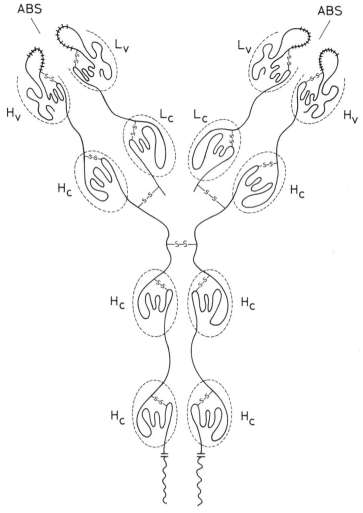

Fig. 4.10. Schematic representation of the structure of immunoglobulin G; H_c and H_v are the constant and the variable domains of the heavy chains; L_c and L_v are the constant and the variable domains of the light chains; the hatched parts of the light and the heavy chains form together the antigen binding sites. ABS is antigen binding site.

Immunoglobulins

The most familiar recognition proteins of the immune system are the immunoglobulins, or antibodies. They come in a great number of slightly different forms. These different forms allow for each molecule the recognition of a specific target pattern. The structure of immunoglobulin G (IgG) is shown schematically in Fig. 4.10. The molecule consists of four polypeptide chains, two nearly identical heavy chains and two nearly identical light chains. The chains are held together by disulfide bonds (see section 3.2) to form a Y-shaped molecule. Both the heavy and the light chains are organized into separately folded domains, four in the heavy chains and two in the light strands.* The structure, therefore may resemble a complex having twelve subunits, even though it actually is a single covalently bound entity.

The domains at the end of both arms of the "Y" (which is the animo end of both the heavy and the light chain) differ from the other domains. It is this difference which is responsible for the wide variability of these molecules. The domains are called *variable domains* in contrast with the other, so called *constant domains.* Within the variable region of the variable domains in each chain there is a small region where the amino acid sequences are particularly diverse. These segments come together at the end of each arm of the Y, forming a cleft that is the antigen-binding site. Thus, each immunoglobulin molecule has two binding sites for the antigen against which it is specific. Because of the spacial flexibility of the Y-shaped molecule it is likely that, once one site is bound the other will also become attached. This gives rise to the typical antigen-antibody precipitate, illustrated in Fig. 4.11, when the antibodies come in contact with antigen.

The specificity of the molecule depends on the shape of the clefts which is determined by the amino acid sequence in the "hyper-variable' segments of the variable domains. Since there are millions, or perhaps billions types

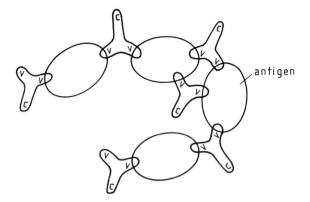

Fig. 4.11. Several immunoglobulin molecules in a typical antigen-antibody precipitate. The ovals represent the antigen.

---

* In some other kinds of immunoglobulins there are five domains in a heavy chain.

of antigen, there must be a way in which the organism is able to synthesize that many antibodies without requiring too much of its genome. The way in which this is accomplished will be discussed in section 4.5.

MHC proteins and receptors

The three-dimensional structure of antibodies was known long before ideas began to develop about the three-dimensional structure of the other immunoproteins. Antibodies appear in blood serum in large and soluble quantities once a *B* cell is confronted with an antigen. *T* cells, however, never differentiate into cells that secrete receptors. Moreover, *MHC* proteins and receptors are very difficult to isolate. Only when it became technically possible to make *T* cell clones or *T* cell hybridomas (a cell fused with another cell) for specific antigens and specific *MHC* proteins that propagate themselves in culture, could some progress be made in the elucidation of the structures.

Figure 4.12 shows schematically what, at the time of this writing, is known about the structures of class I *MHC* protein, class II *MHC* protein and the *T* cell receptor.* The structures show similarities with the immunoglobulin structure. Class I *MHC* protein has a single α chain and occurs always in association with a non-*MHC* protein, β-2-microglobulin. Both class II *MHC* protein and *T* cell receptor have an α chain and a β chain. The chains are made up from domains of some 70 to 100 amino acids held together by a disulfide bond. The amino acid sequences in the domains of the *MHC* proteins are relatively constant within a organism. The *T* cell receptor, like immunoglobulin, has a constant and a variable domain in each chain. The variable domains show a degree of diversity similar to that of immunoglobulin; it is acquired in a similar way (see section 4.5).

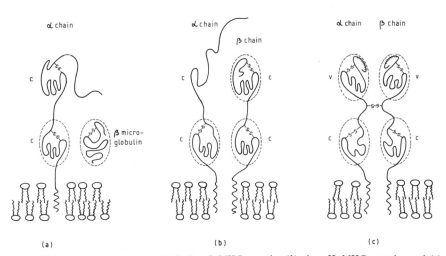

Fig. 4.12. Molecular structures of (*a*) class I *MHC* protein; (*b*) class II *MHC* protein; and (*c*) I cell receptor; c and v are the constant and the variable domains, respectively.

---

* Recently, a second T cell receptor has been identified (Brenner et al: *Nature* **325** (1987) 689).

4.4 The genetic system

The gene
Althoug DNA was known in the nineteenth century, its presence in the chromosomes of a cell was interpreted as a structural necessity. The idea of the gene as the functional unit of heredity already existed, and the distinction between a *genotype* and a *phenotype*, the latter being the manifestation of the function of genes (the so-called genetic expression), was recognized as useful. In the beginning of the twentieth century it was found that the gene itself was composed of DNA and not of protein, as was formerly believed. An important refinement of the concept of genetic expression was the realization that this expression occurs in a very simple way, namely through the production of a unique characteristic protein species. The problem at that stage was the question of how the information contained in a particular DNA molecule could manifest itself in a particular protein molecule. With the realization that DNA as well as protein molecules are linear polymers came the idea that the primary structure (the sequence of monomeric units) in each is all important and that the problem may well turn out to be a coding problem. A major factor for the confirmation of this was, of course, the 1953 determination of a satisfactory steric model of DNA by Watson and Crick and by Franklin and Wilkins. For their interpretation, they made use of the techniques of X-ray diffraction described in section 3.6.

DNA
Watson and Crick proposed a model for the structure of DNA which was made up of two antiparallel intertwisted helices with an overall diameter of about 18 Å. The phosphates in the sugar-phosphate sequence are connected at one side to the fifth carbon atom of the sugar and at the other side to the third carbon atom of the sugar (cf. Fig. 2.9). With respect to the bases, the sequence runs 5'3' in one strand and 3'5' in the other. The strands are made up of sugar-phosphate-sugar-etc., sequences and the bases (attached to the sugars) protrude to the inside. The bases attached to one helical strand are linked to those attached to the other helical strand by hydrogen bonds. This can be accomplished only when there is an exact fit. Measurements based on X-ray diffraction patterns of crystals of the purine and pyrimidine bases have shown that this exact fit exists when adenine (A) is paired with thymine (T) through two hydrogen bonds, and when guanine (G) is paired with cytosine (C) through three hydrogen bonds. As a consequence the relative concentrations of the bases in DNA must satisfy the following relation

$$[A]/[T] = [G]/[C] = 1 .$$

This relation had indeed been verified and was one of the pieces of evidence used by Watson and Crick to construct their model. The DNA molecule is thus like a twisted stepladder with about ten rungs per turn. The rungs, about 11Å long, are formed by the base pairs and linked to each other as depicted in Fig. 4.13.

124

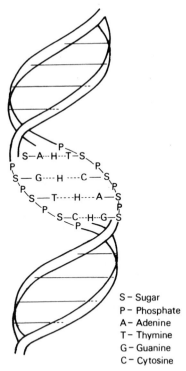

S – Sugar
P – Phosphate
A – Adenine
T – Thymine
G – Guanine
C – Cytosine

Fig. 4.13. The double helix of DNA. The two helically wound strands are bound to each other by hydrogen bonds between complementary bases. The diameter of the spiral is 18 Å and the spacing between turns is 34 Å. The "rungs" between the strands are about 11 Å long and there are about 10 "rungs" per turn. The molecules are very large. The molecular weight of bacterial DNA (*E.coli*) can be as much as 2.8×10^9. Eukaryotic cells contain a species specific number of chromosomes and each chromosome may contain one or more very large DNA molecules (the total length of all the DNA contained in the 46 chromosomes of human cells has been calculated to be about 2 m, which is equivalent to about 5.5×10^9 base pairs).

Because of the specific base pairing, each strand can be a mold or a template to form an exact complementary copy of it, just like a photographic print is made from a negative. In the process of replication (see below) the two strands come apart and each of them serves as a template for the synthesis of a new complementary strand. The new strands then condense with the old ones forming two new molecules of DNA which are exact copies of the one old molecule. In this way the sequence of the bases in the DNA is conserved completely and exactly. In the process of transcription (see below) one of the DNA strands is a template for the synthesis of a complementary copy of RNA (with the one variation that in RNA the thymine is replaced by uracyl). This RNA molecule is then used to program the synthesis of protein.

The genetic code
It soon became obvious that the linear sequence of the bases G, C, A, and T in DNA must contain the order for the protein programming, and a great

125

deal of effort then went into deciphering the code. In the years following the discovery of Watson and Crick most of this effort was theoretical. If a particular sequence of two of four bases were assigned to a particular amino acid, only 16 (=4^2) amino acids could be accounted for. The code "word" therefore must have at least three "letters" (a "triplet"). But that gives 64 (=4^3) possibilities and there are only 20 amino acids in the proteins of a living cell. This had been a problem in the early days because it seemed to be difficult, at that time, to think of a degenerate code (a code in which more triplets could be assigned to one amino acid). Hence, much effort went into finding groups among the 64 combinations which would turn out the magic number 20. The most brilliant among these efforts was the one of George Gamow who found "diamond-shaped" pockets in the Watson-Crick double helical model. These pockets were formed and bounded by a base on one DNA chain, the adjacent base pair (spanning both chains), and the next adjacent base beyond the pair on the other chain. The 64 possible diamond-shaped pockets could be divided into 20 categories each of which contain diamonds that do not change in character. The residues of the 20 kinds of amino acids would each be a correct fit for one of the corresponding categories of diamond-shaped pockets.

There are some elements in this hypothesis which have turned out to be correct. The code is indeed a "three-letter" code (in Gamov's model each of the two bases in the base pair of the boundary of the pocket is, by the complementary C-G and A-T binding, completely determined by the other so that there are only three "independent" bases in the boundaries of the pockets) and there is a certain degree of degeneracy. The most important thing which is preserved is the template idea; in Gamow's model the DNA molecule serves as a template for the peptide chain.

What turned out to be wrong in Gamow's idea is that, in his view, the DNA itself is the template for the protein synthesis. The idea of an intermediate between the DNA and the site of protein synthesis is largely owed to Jacob and Monod (1961), who defined a "messenger". Confirmed by a wealth of experimental evidence, it turned out that a special kind of RNA, the *messenger RNA* (mRNA), serves as such a messenger (see below).

As remarked, much of the effort to assign particular codons to particular amino acids (the deciphering of the code) had, in the earlier days, been theoretical. The results were, consequently, of a rather speculative character. This changed radically when Nirenberg and Matthaei (1961) isolated an *in vitro* protein synthesizing system through which an experimental approach to the problem became possible. By letting protein synthesis be programmed by synthetic RNAs (such as the polynucleotide poly(U), or copolymers of repeating sequences, UCAUCAUCA ..., or other variations) the assignments could be made within a few years.

The genetic dictionary is now known beyond much doubt (see Table 4.1). The codons UAA, UAG, and UGA are involved in peptide chain termination punctuation. Further, the code is highly degenerate; in a very general way those amino acids which occur with high (or low) frequency in protein tend

126

Table 4.1. Codon assignements.

	U	C	A	G	
	phe	ser	tyr	cys	U
U	phe	ser	tyr	cys	C
	leu	ser	stop	stop	A
	leu	ser	stop	trp	G
	leu	pro	his	arg	U
C	leu	pro	his	arg	C
	leu	pro	gln	arg	A
	leu	pro	gln	arg	G
	ile	thr	asn	ser	U
A	ile	thr	asn	ser	C
	ile	thr	lys	arg	A
	met	thr	lys	arg	G
	val	ala	asp	gly	U
G	val	ala	asp	gly	C
	val	ala	glu	gly	A
	val	ala	glu	gly	G

to have a large (or small) number of codons assigned to them. The universality of the code is remarkable. RNAs with a specific code give rise to the same polypeptide chains, irrespective of whether the protein synthesizing system originated from a bacterium or a eukaryotic cell from a higher organism.

RNA

Three kinds of RNA are directly involved in the synthesis of proteins which must have the correct order of amino acids dictated by the nucleotide sequences in DNA: *Ribosomal* RNA (rRNA) which is an inherent factor of the ribosomes, the sites of protein synthesis; *messenger* RNA (mRNA) that conveys the information expressed in the DNA nucleotide sequences to the sites of protein synthesis; and *transfer* RNA (tRNA) which serves as the adaptor between the mRNA and a particular amino acid. In addition a cell can contain smaller RNA types which perform a variety of regulatory functions. All types of RNA originate from DNA transcripts.

rRNA is part of the ribosomes, the small particulate organelles which are the site of protein synthesis. A ribosome consists of a large (50 S) and a small (30 S) ribonucleoprotein particle, each of which is a complex of some protein molecules bound to rRNA. The large particle of a eukaryotic ribosome contains three rRNA molecules, a long one of about 4,500 nucleotides and two shorter ones of about 160 and 120 nucleotides. The small particle contains a single long rRNA of about 1,800 nucleotides. This composition helps to give the ribosome a grooved structure which can accomodate both a mRNA and a protein at the same time. In eukaryotes as well as in prokaryotes a single rRNA transcript is cut in several places to form the major types of

127

rRNA, which are then bound to protein molecules to form the two ribosomal subunits. In eukaryotes these processes take place in the cell nucleus; both subunits are transported out of the nucleus through pores in the nuclear membrane and assembled in the cytoplasm in close association with the endoplasmic reticulum. In prokaryotes, which have no nucleus, the assembly of the ribosomal subunits takes place in the cytoplasm together with the binding of mRNA and tRNA (see below).

tRNA serves as the adaptor between the mRNA and a particular amino acid. It is a smaller molecule, about 70 to 80 nucleotides long. There seem to be about 40 different species. Some of the bases of the tRNA, such as the base *pseudouridine* (designated by ψ), deviate from the usual RNA bases A,U,C and G. Pseudouridine is common to all normal tRNAs. It resembles uracil but it is attached to the ribose at the 5th (C) position of the ring rather than at the 1st (N) position (see Chapter 2, section 2.3). Based on sequencing, the secondary structure is believed to follow a looped, so-called cloverleaf pattern, generated by base pairing at several places of the molecule. This cloverleaf pattern is shown in Fig. 4.14a. The long stem at the top, called the *acceptor stem* contains the site for the attachment of the amino acid specific for that tRNA. The loop at the bottom has a sequence of three specific bases, forming the *anticodon*; the anticodon hydrogen-binds to the complementary codon of the mRNA (see below). There is still no clue as to how a tRNA with a specific anticodon at its bottom loop recognizes the compatible amino acid for binding at its acceptor end.

The tertiary structure of phenylalanine-tRNA, revealed by X-ray diffraction studies carried out by the group of Alexander Rich at MIT is shown in Fig. 4.14b. This structure, which is generated by additional, so called tertiary hydrogen bonds, is fully compatible with the cloverleaf secondary structure illustrated in Fig. 4.14a.

Also tRNA is transcribed from DNA and then processed. The ends of the primary transcript are cut and the cloverleaf structure is formed by base pairing. In eukaryotes this process takes place in the nucleus and the completed tRNA is transported to the cytoplasm.

mRNA is the conveyer of the information for protein synthesis. As can be understood from its function, mRNA is heterogeneous; it provides templates for some 4000 different kinds of proteins in a "typical" cell. A typical polypeptide of 40,000 daltons requires a corresponding messenger RNA of 400,000 daltons, or a "tape" length of 1200 nucleotides. It is generated by a process called transcription (see below) either directly (in prokaryotes) or after processing (in eukaryotes).

The tape reading processes

The transformation of genetic information can essentially be seen as occurring in three more or less complicated tape reading processes (see Fig. 4.15). First, there is the process of *replication* in which a DNA input tape is read out by "tape readers", a group of enzymes called *DNA polymerases*. DNA polymerase III catalyzes a complete unwinding of the DNA double helix and

128

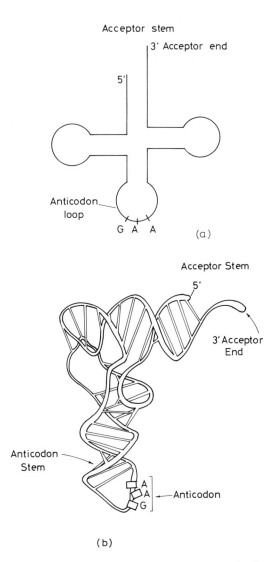

Acceptor stem

3' Acceptor end

5'

Anticodon
loop

G A A   (a)

Acceptor Stem

5'

3' Acceptor
End

Anticodon
Stem

A
A
A
G

Anticodon

(b)

Fig. 4.14. (*a*) The "cloverleaf" secondary structure of tRNA; the anticodon corresponds to the amino acid phenylanaline. (*b*) The tertiary structure of phenylanaline tRNA (from G.J. Quigley and A. Rich, Science **194** (1976) 796-806.

a simultaneous synthesis of two DNA strands, one in the opposite direction from the other. The outputs are two double helical molecules of DNA. In each of the new molecules one of the strands came from the original DNA molecule; the other strand is synthesized on the original one, their bases being complementary and hydrogen bonded to the bases of the original strand. DNA polymerase I, re-reads the newly formed strand and corrects errors. The tape reading process thus turns out two new DNA tapes identical to the original one. Each DNA tape is read only once during the cell cycle.

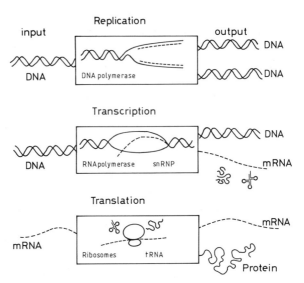

Fig. 4.15. The three tape reading processes of a cell.

In prokaryotes, the time it takes is almost equal to the doubling time; DNA synthesis, hence takes almost the complete life cycle of the cell. In eukaryotes DNA replication takes place only during a specific phase (the S phase) of the life cycle. What triggers the initiation of the process is still unknown. The two new DNA tapes become the complete "library", called *genome*, for the daughter cells after cell division.

Second, there is the process of *transcription*. In this process parts of the DNA tape are read by transcription tape readers (RNA polymerases). In prokaryotes a single RNA polymerase catalyzes all RNA synthesis. In eukaryotes, however, three distinct types of RNA polymerase are active; type I transcribes the precursor of rRNA, type II the precursor of mRNA and type III the precursor of tRNA and other small RNA's. The polymerase recognize and bind to a region of the DNA double helix characterized by a special sequence of nucleotides. This region is called the *promotor*. It is situated just outside the beginning of each gene (or cluster of genes) that is to be transcribed. The enzyme unwinds part of the double helix exposing the two strands of DNA, one of which is then transcribed. The transcription is the binding of ribose nucleotides with bases complementary to those of the DNA strand as the polymerase slides along the DNA. Each incoming ribose nucleotide has a triphosphate group at its 5' position (see section 2.3). A pyrophosphate is lost and the remaining phosphate makes a phosphodiester bond with the hydroxyl (OH) group at the 3' position of the nucleotide at the end of the chain. Thus the transcription goes "downstream" in the 5' to 3' direction. It goes on until a termination codon (see Table 4.1) is passed.

Finally, there is the process of *translation*. Each mRNA carries the information contained in the sequence of its nucleotides (or more precisely in the sequence of its codons, "words" of three nucleotide "letters" each) to the

130

translation machines (the ribosomes and the "charged" tRNA's) where the codons are translated into amino acids which then are incorporated into the peptide chain in the order prescribed by the order of the codons in the mRNA. Due to a recognition process which, as already mentioned, still is not understood, the tRNA, through a charging process involving the energy carrier adenosine triphosphate (ATP, see Chapter 5), binds with its assigned amino acid. The so-formed aminoacyl-tRNA can be attached to a ribosome when the latter is connected to part of the mRNA single strand (see Fig. 4.16). In this process the anticodon of the tRNA is bonded (by hydrogen bonds) to the appropriate codon of the mRNA. The adjacent codon of the mRNA can then bind the appropriate aminoacyl-tRNA through its anticodon and the two amino acids form a peptide bond when the correct factors are present (among others K^+ ion and guanosine triphosphate). The ribosome then moves one position further along the mRNA, the "uncharged" tRNA loosens itself from the ribosome, and the next aminoacyl-tRNA can attach itself, thus placing its amino acid in the right position. The exact mechanism of this process is not yet known but it is clearly demonstrated that the role of the tRNAs and the mRNA in the ribosome is, indeed, as described. The translation machine thus forms a polypeptide chain (a protein) exactly in the order as prescribed by the mRNA.

4.5 Regulation and control

Regulatory proteins
It is obvious that the tape reading processes in a cell must be regulated and controlled. In every kind of cell decisions must be made as to which genes are to be transcribed into mRNA for subsequent translation into protein and

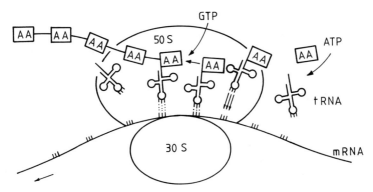

Fig. 4.16. The translating machinery of the ribosome. Specific transfer-RNA molecules form complexes each with a specific amino acid. The anticodons of these aminoacyl-tRNAs form hydrogen bonds with complementary codons of the mRNA which is attached to the ribosome. The amino acids, thus placed in a sequence dictated by the codon sequence of the mRNA, form peptide bonds after which the tRNAs detach.

when. Much of this control is exerted at the level of the binding of RNA polymerase to the promotor region of DNA. It is the tightness of this binding which turns the transcription tape reading on and off. One can easily imagine that this binding can be inhibited or stimulated by other regulatory proteins that either block the polymerase binding site or change the conformation near the binding site in such a way that close fit binding is promoted. Such kinds of mechanisms of control indeed occur. It appears that binding of RNA polymerase is regulated by many kinds of special nucleotide sequences on the DNA and many proteins that recognize them.

The first evidence for such regulatory proteins came from the study of bacteria and viruses. In the colon bacterium *Escherichia coli*, for example, the metabolism of the sugar lactose depends on the concentration of lactose present in a cell. The first enzyme which is involved in this metabolism is β-galactosidase which splits lactose into two smaller sugars. If the bacteria are grown in a medium in which no lactose is present, no synthesis of the enzyme takes

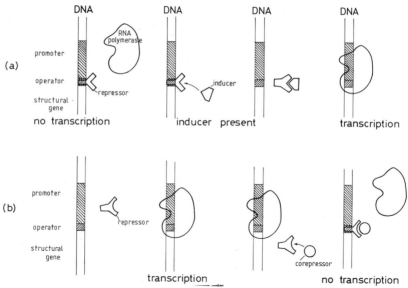

Fig. 4.17. (*a*) The mechanism of induced enzyme synthesis. The repressor molecule has a binding site specific for a binding site on the operator. When the repressor is bound to the operator the structural gene cannot be transcribed. The inducer (for instance lactose) can bind to a complementary binding site on the repressor. When the inducer is bound to the repressor, the conformation of the latter is altered to such an extent that it detaches itself from the operator. The structural gene is then free to be expressed and enzyme (for instance β-galactosidase) synthesis can proceed.

(*b*) The mechanism of repressed enzyme synthesis. The repressor molecule has a conformation which is not compatible with the binding site on the operator. When a corepressor (for instance histidine) binds to a complementary binding site on the repressor, the conformation of the latter becomes compatible with the binding site on the operator. The repressor-corepressor complex binds itself to the operator and the expression of the structural gene (the synthesis of, for instance, hisdehydrogenase) is blocked.

132

place. However, as soon as lactose is added to the growth medium, synthesis of β-galactosidase is turned on and proceeds at a high rate.

This "turning on and off" (see Fig. 4.17) is accomplished by a regulatory protein, called *repressor* which can bind to a sequence of DNA nucleotides situated between the promotor and the gene (or gene cluster) that codes for the enzyme (or group of enzymes). Such a gene or gene cluster is called a *structural gene or structural gene cluster*, in order to distinguish it from genes that code for regulatory proteins. The DNA sequence to which the repressor binds is called *operator*. Often the operator overlaps partly with the promotor. Binding of the repressor prevents the RNA polymerase to bind to the promotor so that the transcription cannot take place. The binding of the repressor to the operator depends on an *effector*; the binding of such an effector to the repressor protein determines whether or not the repressor can bind to the operator. In the case of lactose metabolism the effector molecule is lactose; in the absence of lactose the repressor binds to the operator and no transcription can take place. If lactose is present however it binds to the repressor; the repressor undergoes a conformational change and is released from the operator. The polymerase now binds to the promotor and transcription of the β-galactosidase gene is induced. The effector in this case is an *inducer*.

Some effector molecules act the opposite way. Then they are *corepressors*. In that case the repressor binds tightly to the operator only if it is bound to the corepressor. An example is the synthesis of the amino acid histidine. There are a number of enzymes involved in this synthesis, one of them being histidinol dehydrogenase. Transcription of the gene coding for this enzyme is inhibited in the presence of histidine which acts as a corepressor molecule.

Repression and induction of enzyme synthesis often occurs in coordination by a single corepressor or inducer. In many cases a product or substrate is involved in a sequence of enzymatic reactions in which the enzymes are related to each other. Those enzymes are coded by structural genes that are often (but not always) adjacent to each other in the DNA. Together they constitute an *operon*; they are turned on an off by a single repressor. For instance, the synthesis of histidine proceeds from phosphoribosyl pyrophosphate in ten consecutive steps involving nine different enzymes. The structural genes for these enzymes are adjacent to each other and constitute together the *his operon*. The transcription of the entire operon is repressed when histidine binds with a single repressor and the consequent binding of the repressor-histidine complex to a single operator. This also applies to the *lac operon* which has three structural genes for the three enzymes involved in lactose metabolism.

Processing and control in eukaryotes
In prokaryotes which do not have a nucleus, protein synthesis starts immediately; the transcribed mRNA is engaged by the "waiting" ribosomes and tRNAs even before its synthesis has been completed. Transcription in eukaryotes however, is followed by sometimes substantial processing of the primary transcript RNA. This is necessary because in nuclear DNA the nucleotide

sequences that code for a protein are spread out over the genome and appear in pieces that often are separated over substantial distances by intervening non-coding sequences. The pieces of coding sequences are called *exons*; the intervening sequences that contain no relevant coding information are *introns*. The introns must be excised and the exons rejoined, or spliced first in order to make the mRNA. This process provides for an efficient gene regulation.

The primary transcript, which is often called *heterogeneous nuclear* RNA, or hnRNA, thus undergoes substantial processing before it leaves the nucleus as mRNA. During the transcription process it is provided by a chemically protective "cap" (a methylated guanosine, which is a guanine base attached to a ribose) at its 5' end and a tail (a series of between 150 and 200 adenine nucleotides) at its 3' end. Parts of the transcripts are cut out from the middle and the ends of the remaining parts are spliced together. This cutting and splicing is under enzymatic control; the enzymes operating the cutting belong to a class of small ribonucleoproteins (snRNP). The resulting mRNA thus is shortened but has retained its cap and tail. The cut-out parts are transcripts of the introns mentioned above. What remains on the mRNA are the transcriptions of the DNA exons (see Fig. 4.18). In many genes, such as those coding for hemoglobin, antibodies and a number of enzymes these exons seem to encode for domains of the protein that are recognizable as functional units.

It is clear that this processing provides for control of gene expression as well as for variability of the system. It could play a role in the differentiation of cells in the higher organism. Also, different mRNA's coding for different proteins can be made from the same primary transcript. Examples of this differential processing are known.

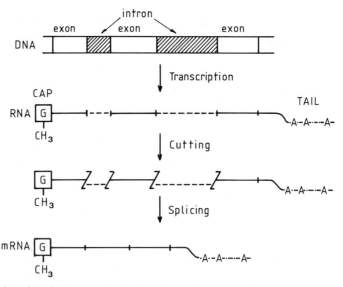

Fig. 4.18. Processing of the RNA transcript in the nucleus of an eukaryotic cell.

134

Structural variations of DNA

Polymerases and regulatory proteins bind specifically to the promotor, respectively operator regions on the DNA. These regions contain specific nucleotide sequences that are recognized by the proteins. The proteins bind to the recognized sites by forming hydrogen bonds linking a defined set of amino acid side chains to a corresponding set of hydrogen-bond donors and acceptors on the specific DNA regions. The recognition process is largely facilitated by the fact that the base sequences modify locally the double helix structure of the DNA.

Presently one recognizes three different DNA structures. The structure that was proposed by Watson and Crick is now called the B-structure of DNA. It is characterized by a major groove adjacent to a minor groove with approximately equal depth along the DNA helix. Runglike base pairs are sitting astride the helical axis and are perpendicular to it. In DNA having an A-structure, the base pairs are tilted and are pulled away from the axis of the double helix; the major groove has become far deeper than the minor groove. The Z-structure, finally, is more or less the inverse of the A-structure. Whereas both A-DNA and B-DNA are right-handed helices, Z-DNA is left-handed with a deep minor groove and a flattened major groove.

There is recent evidence that these three structural forms occur locally and that such local variations are caused by sequences of interacting base pairs. The regulatory proteins, therefore, recognize the regulatory DNA regions by shape fitting that allow close ranges for forming the specific hydrogen bonds. The evidence was substantially supported by information obtained from X-ray diffraction studies carried out recently in the laboratory of B. Matthews of a complex of DNA and a repressor molecule in a bacterial virus. The repressor was the *cro* repressor that takes part in a regulatory process somewhat like that of the *lac* repressor but more complex. Thus the nucleotide sequence in DNA not only encodes for the amino acid sequence of proteins but also for the instructions for the selective expression of the genes.

Somatic recombination

From what has been discussed so far one should not conclude that the DNA itself always plays the passive role of an archive. For instance, the application of recently developed recombinant DNA technology (see section 4.6) to the study of immunoglobulin genes has shown that in mature antibody-secreting cells parts of the DNA have been reshuffled and recombined. It appeared that in embryonic cells the genes for antibodies are broken up into small segments that are widely scattered throughout the genome. As the *B* cell matures, specialized restriction enzymes, called *endonucleases*, cut the DNA at recognized positions. Other enzymes, called *ligases*, join the selected pieces together.

Such somatic DNA recombination processes may turn out to be a major factor for the transcription of antibody genes. It appeared that the genes that encode for the different domains of the immunoglobulin molecule, and especially those for the hypervariable domains are widely scattered throughout the genome in the germ-line cells that eventually differentiate into *B* cells. During the differentiation process, some of the genes are cut out randomly and a new antibody gene is assembled in the mature *B* cell. The antibody then is synthesized after transcription and processing of the RNA as described above. In this way a wide diversity of antibodies can be obtained in an organism.

4.6 Recombinant DNA

Restriction enzymes

In spite of many advances made by research during the two decades following the discovery of the structure of DNA one substantial problem remained. That problem was the inaccessibility of DNA. The localization and analysis of specific DNA sequences was possible only indirectly by looking at base sequences in mRNA and amino acid sequences in protein. This situation changed dramatically with the discovery of *restriction endonucleases* in the mid-1970s.* These DNA cleaving enzymes cut a DNA molecule only at specific sequences that occur here and there along the DNA double helix. Several hundreds of them are now known and available.

An example of a restriction endonuclease is *EcoRI* which is found in the colon bacterium *E. coli*. This enzyme cleaves DNA when it recognizes the sequence GAATTC in the 5′→3′ direction. This sequence is double symmetric; it is similar in both complimentary strands when read in the same direction. The cleavage occurs between the first G and A in both strands so that the termini at both segments have a few single-stranded sequences:

$$5'....G{\downarrow}A-A-T-T-C....3'$$
$$3'....C-T-T-A-A{\uparrow}G....5'$$

$$5'...G\ 3' \qquad\qquad \Big\downarrow \qquad\qquad 5'A-A-T-T-C....3'$$
$$3'....C-T-T-A-A\ 5' \qquad\qquad\qquad\qquad 3'G....5'$$

Many other restriction endonucleases act the same way but, of course, at different sequences. Some others cleave DNA in fragments with complete base-paired termini.

Restriction enzymes have become powerful tools, not only for experimenters but also, as will become clear, for use in biotechnology. With them it became possible to reduce the very long DNA molecule into sets of discrete fragments that can be separated by gel electrophoresis. Each fragment can then be

---

* The term "Restriction Enzyme" was coined in the mid 1950's when it was discovered that some bacterial strains can break down phages which were grown in other strains. Growth of the phage was inhibited or "restricted". When the same phage was grown in the original strain, no restriction of phage growth occurred. It appeared later that an enzyme was responsible for the restriction.

subjected to further analysis. Also, fragments can be joined together by mixing them in the presence of DNA ligase. The analysis of the fragments became relatively easy when several procedures for the direct sequencing of DNA were developed. Such procedures now allow remarkably rapid determinations of the entire base sequence of segments generated by restriction enzyme cleavage. Because the genetic dictionary (the translation of a base sequence into an amino acid sequence) is known, the primary structure of proteins can be deduced from its DNA sequence and genes can be, and were, recognized in the DNA fragments. Moreover, sequences that are not involved in encoding proteins but in regulating the expression of genes (see previous section) could be determined.

These developments have led to a set of procedures, used in research as well as in technological applications, known as *Recombinant-DNA Technology*. Ultimately these procedures are based on the similarity of the molecular organization in all organisms, from viruses and bacteria through mammals. DNA, whatever its origin, is structurally compatible with DNA of any other origin. As a result, DNA segments from one form of life can readily be blended with DNA from whatever other form.

Phages and plasmids as vectors
Because of this structural compatibility *phages*, which are viruses that infect bacteria, can inject their DNA into a bacterial cell and cause it to be replicated many times over, package the newly synthesized DNA into viral protein coats and kill the cell. Other more friendly bacterial parasites are the *plasmids*, which are circular pieces of DNA growing inside the bacterial cell. Because plasmids can carry genes that confer certain advantages to their host (such as resistance to an antibiotic) the cell allows the plasmid DNA to be replicated and to be transferred to the daughter cells when the bacterium divides.

Some phage and plasmid DNA's are small in size and, therefore easy to manipulate and restructure by following recently developed procedures. They can be isolated and cut at specific sites with restriction enzymes. The resulting fractions can be rejoined (to reconstitute the original molecule) or joined to foreign DNA (to make a hybrid molecule). The joining is easily done by mixing the fragments with DNA ligase. A hybrid molecule made of a plasmid fused with a foreign DNA segment can be reintroduced into a bacterial cell. The plasmid, including the foreign segment then is replicated. The plasmid thus is a carrier of foreign DNA, establishing and "amplifying" it in bacteria. Such a carrier is called a *vector*. Also phages can serve as vectors; they can be used to convey the foreign DNA from one bacterium to another.

Cloning
If a DNA segment from, say, a mammalian genome is inserted in a vector and introduced into a bacterial cell, all the descendants of this bacterial cell will contain the same segment. The segment is multiplied and reproduced in its pure original form. The population of these similar DNA segments (of course still contained in the vector) is a *clone* of the original DNA segment.

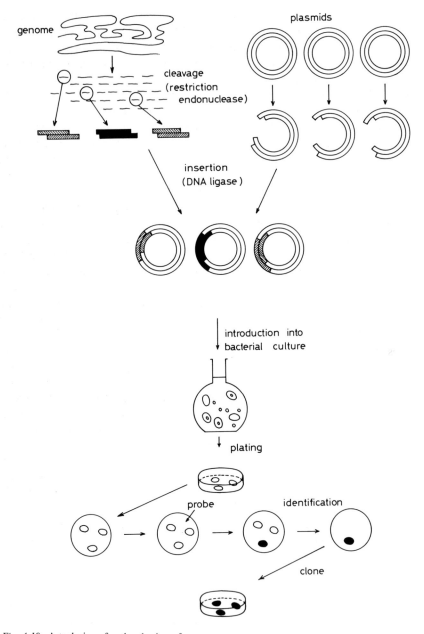

Fig. 4.19. A technique for the cloning of a gene.

The process of making such clones is central to the operation of recombinant-DNA technology.

Cloning starts with whole cellular DNA of some organism (see Fig. 4.19). The DNA is cleaved by a restriction enzyme into fragments of a size that can be accomodated by the carrying capacity of a vector, a plasmid for example.

A mammalian genome, as for instance the human one, can be broken up into a few hundred thousand fragments. When these fragments are mixed together with a large number of plasmids that are cleaved with the same restriction enzyme, the insertion of the fragments into the vector occurs in minutes after the addition of bacterial DNA ligase. The sealing is largely facilitated when a restriction enzyme, such as EcoRI is used that leaves single stranded termini, so called "sticky ends", at the fragments. The hybrid molecules, part plasmid and part mammalian DNA, then are introduced into bacterial cells, where they are replicated. Each mammalian segment is replicated in the progeny of a single bacterium. The clones can be singled out by plating the bacteria on a culture disk, diluted in such a way that the resulting colonies, each descending from a single cell, are physically separated.

In such a way the insertion and replication process gives rise to an array of sometimes hundreds of thousands of distinct and physically separated clonal populations. Such an array is called the *library* of the genome. Of course, specific clones of such a library still need to be identified. Several procedures have been developed for this purpose. Most of such procedures make use of specific "probes". If a gene or a DNA segment has been cloned before, this DNA can serve as a probe when it is labeled with a radioactive isotope (such as the radioactive phosphorus isotope ^{32}P). The radioactive DNA forms a hybrid with the complementary strands of the clonal DNA. One can take samples from the library, break the cells open to expose the DNA, add the probe and expose the sample to a photographic plate, thus identifying the clone.

To isolate genes that have never been cloned before one must find other kinds of specific probes that scan the library and identify the clone of interest. Many such probes have already been found and succesfully used. Some of these involve mRNA's which are isolated from differentiated cells that are specialized to make a specific protein in large quantities. An example is the protein *globin*, a precursor of hemoglobin, that is produced in quantities far exceeding that of other proteins in red blood cells. Other probes have been developed by starting from a protein, the amino acid sequence of which is (partly) known. Small DNA molecules then can be synthesized from "off-the-shelf" nucleotides, back-translating the amino acid sequence into nucleotide sequence. In principle, any gene whose protein product is known can now be isolated.

Applications
When a gene is isolated in sufficient quantities by the cloning procedure described above it can be inserted into a foreign cell, usually a bacterial cell, which can be forced to express it. The gene of interest is excised from the vector in which it was cloned and then modified. This modification is necessary because the mammalian gene bears regulatory sequences that promotes transcription in a mammalian cell and not in a bacterial cell. Such regulatory sequences have to be replaced by bacterial regulatory sequences. The modified gene is inserted into a so called *expression vector*, a plasmid having sequences that facilitates the expression of the gene in a foreign cell. The vector containing

139

the mammalian gene (or a similarly processed plant gene) is then introduced into a foreign cell, usually a bacterial cell that can be easily grown in large cultures.

The application of such procedures on a large scale is the basis of present-day biotechnology. A protein that can be produced only in limited amounts in a mammalian cell now can be synthesized in large quantities in bacterial cultures, rapidly growing in industrial fermentors. An increasing number of useful protein products (such as insulin, growth hormone, the enzyme rennin for cheese making and many others) can be manifactured in a relatively cheap and economic way.

Cloned genes can also be inserted into the genome of an intact multicellular organism. If the gene insertion is done in somatic cells (cells that are not reproduced), it changes the behavior of only those cells. Of greater importance is, of course, the insertion into germ-line cells (such as sperm and egg cells) so that the modified genome has its influence on the entire organism and is also transferred to the descendants of that organism. This would open up far reaching possibilites, such as changing and perhaps improving genetic characteristics of crop and cattle or correcting genetically determined diseases. Techniques for achieving germ-line insertion are indeed now available, although they are still limited in important ways. The insertion, usually into embryonic cells, cannot be directed and integrated into a particular chromosomal site; the regulation of the expression of the inserted gene can therefore be very different from that of the original resident genes. Such limitations presently are very difficult to (and perhaps may never be) overcome.

Bibliography

Alberts, B., Brags, D., Lewis, J., Raff, M., Roberts, D.T., and Watson, J.D. (1983). "Molecular Biology of the Cell", Garland Publ. Cie., New York, London.

Amit, A.G., Mariuzza, R.A., Phillips, S.E.V., and Poljak, R.J., Three-dimensional Structure of an Antigen-Antibody Complex at 2.8 Å Resolution, Science 233, 747–753.

Axel, R., Maniakis, T., and Fox, C.F., eds. (1979) "Eukaryotic Gene Regulation". Acad. Press, New York.

Cantor, C.R., and Schimmel, P.R. (1980). op cit., Chapter 2.

Cantor, C.R., and Schimmel, P.R. (1980). "Biophysical Chemistry Part III: The Behavior of Biological Macromolecules". W.H. Freeman and Cie. San Francisco.

Chambon, P. (1981). Split Genes, Sci. Am. 244 (May), 48–59.

Darnell, Jr., J.E. (1985). RNA, Sci. Am. 253 (October), 54–64.

Doolittle, R.F. (1985). Proteins, Sci. Am. 253 (October), 74–83.

Felsenfeld, G. (1985). DNA, Sci. Am. 253 (October), 44–53.

Gamov, G. (1955). Information Transfer in the Living Cell, Sci. Am. 193 (October), 70–84.

Gothia, C. (1984). Principles that Determine the Structure of Proteins, Ann. Rev. Biochem. 53, 537–572.

Jacob, F., and Monod, M. (1961). Genetic Regulation Mechanisms in the Synthesis of Protein, J. Mol. Biol. 8, 318–356.

McClarin, J.A., Frederick, C.A., Bi-Cheng Wang, Green, P., Boyer, H.W., Grable, J., and Rosenburg, J.M. (1986), Structure of DNA-EcoEndonuclease Recognition Complex at 3 Å Resolution, Science 234, 1526–1541.

Nirenberg, M.W., and Mathaei, H. (1961). The Dependence of Cell-free Protein Synthesis in *E. coli* upon Natural Occurring or Synthetic Polynucleotides, *Proc. Natl. Acad. Sci. U.S.* **47**, 1588–1602.

Pauling, L., and Corey, R.B. (1951). Configuration of Polypeptide Chains, *Nature* **168**, 550–551.

Tonegawa, S. (1985). The Molecules of the Immune System, *Sci. Am.* **253** (October), 104–113.

Watson, J.D., and Crick, F.H.C. (1953). A Structure for Deoxyribose Nucleic Acid, *Nature* **171**, 737–738.

Watson, J.D., Tooze, J., and Kurz, D.T. (1983). "Recombinant DNA". Scientific American Books Inc., New York.

141

5. Biological energy conversion

5.1. The biological energy flow

Biological cycle

Work is continuously being performed by a living organism. Living cells do work either within themselves or on their environment. This work may be of different kinds; mechanical work of muscle contraction, electrical work when charges are transported, osmotic work when material is transported across semipermeable barriers, or chemical work when new material is synthesized. In order to perform all this work the cell needs mechanisms by which energy can be transformed (converted) in the proper way. At the constant temperature prevailing in most cells a net output of work can be obtained only when energy is *dissipated*, that is, when it is converted to a less useful form. The living cell has complex and very efficient devices at its disposal to accomplish this.

The ultimate source of energy for life is the sun. Green plants, algae, and a few types of bacteria are able to capture energy from sunlight and convert it into a form suitable for sustaining their own life and that of the rest of the living world. The process by which this conversion occurs in known as *photosynthesis*. The product of photosynthesis, a large amount of chemical potential energy (food) is then used in a "reverse" process, yielding the form of energy suitable for the performance of work (Fig. 5.1).

Photosynthesis

In photosynthesis, the excitation of a specialized kind of molecules in the so-called *reaction centers* leads to a primary oxidation-reduction reaction which sets in motion a sequence of reactions of the same type (see Chapter 6). The final result of this is the oxidation of a hydrogen donor H_2A and the concomitant production of a relatively strong reductant H_2X:

$$H_2A + X + light \rightarrow H_2X + A. \tag{5.1}$$

The reductant is then used to reduce carbon dioxide to a sugar,

$$12H_2X + 6CO_2 \rightarrow (HCOH)_6 + 6H_2O + 12X \tag{5.2}$$

in a reaction which does not need light. The overall reaction of photosynthesis can, thus, be written as

$$12H_2A + 6CO_2 + light \rightarrow (HCOH)_6 + 6H_2O + 12A. \tag{5.3}$$

143

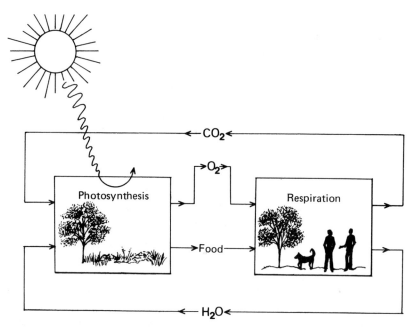

Fig. 5.1. The biological energy cycle. The energy of sunlight is captured by photosynthesis in the form of the chemical potential energy of food and oxygen. The respiratory processes recombine the two to form the lower energy products carbon dioxide and water. The energy liberated in this way is used to preserve and reproduce life. Carbon dioxide and water are again used to carry out the photosynthetic reactions, capturing sunlight, and the cycle is repeated.

In photosynthetic bacteria, A can be a variety of substances ranging from sulfur to organic groups. In higher plants and algae, however, A is always oxygen so that the donor is always water. The overall reaction for these organisms thus becomes

$$6H_2O + 6CO_2 + light \rightarrow (HCOH)_6 + 6O_2 \qquad (5.4)$$

(the form in which it is generally known). The energy term in the left side of this equation (light) is the electromagnetic energy of absorbed light. Energy is represented in the right-hand side of the equation by the chemical potential energy in the substances $(HCOH)_6$ and O_2. One can visualize this by realizing that the compounds on the left-hand side of the equation, water and carbon dioxide, are more stable than the sugar and oxygen on the right-hand side; the total binding energy on the left-hand side is lower (more negative) than that of the right-hand side. Part of this difference is necessarily lost because the reaction is not reversible. The rest is the energy stored as chemical potential energy in the sugar and oxygen.

Respiration

Recombination of the sugar and the oxygen can lead to the liberation of this stored energy. If this recombination occurs by combustion in a test tube all the energy is liberated in the form of heat which as such cannot be used

144

to do work at the constant temperature prevailing in most organisms. When this release of energy occurs, however, by a balanced sequence of oxidation-reduction reactions such as those occurring in the mitochondria of living cells, the energy is liberated in a stepwise manner and caught in a form (again as chemical potential energy) which makes it possible to perform work at the necessary time and place. This form of energy capture occurs in the compound *adenosine triphosphate* (universally known by its initials ATP) which is produced during the the stepwise oxidation of food. The hydrolysis reaction

$$ATP + H_2O \rightarrow ADP + P_i \qquad (5.5)$$

in which ADP stands for adenosine diphosphate and P_i for inorganic phosphate, occurs with a release of energy which, when the reaction proceeds under controlled conditions, can be used for work. The controlled progress of the biological energy cycle can be depicted by the running of a series of water mills driving generators which charge batteries (Fig. 5.2). When photosynthesis is compared to a pump used to bring "water" to an elevated level "using electromagnetic energy", respiration can be represented as the stepwise downfall of the "water" which drives the "water mills" charging the "ATP batteries". The batteries then can be transported to sites where work has to be done; when properly connected, they can be "discharged" by the hydrolysis reaction when work is performed.

Although the coupling of energy-releasing oxidation reactions and ATP formation, as well as the hydrolysis of ATP during the performance of chemical, electrical, osmotic, or mechanical work, may be thermodynamically simple,

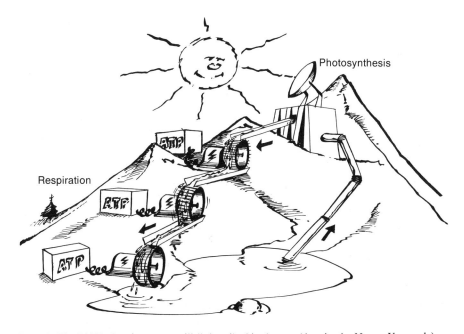

Fig. 5.2. The "ATP-charging water mills" described in the text (drawing by N. van Yperzeele).

145

its molecular mechanism (or mechanisms) is not yet fully understood. It still constitutes a central and challenging problem of present research in bioenergetics. We will discuss it in more detail in this and the following chapters.

5.2 Adenosine triphosphate in coupled reactions; pyridine nucleotides

High energy phosphates
ATP, the universal energy currency in living systems, is a nucleotide. It consists of a base, adenine, of the purine group which is linked by a glycosidic linkage to a molecule of D-ribose. A row of three phosphate groups is attached to the 5'-position of the ribose (see Fig. 5.3). If the terminal phosphate group is detached the nucleotide becomes adenosine diphosphate (ADP). With only one phosphate group we have adenosine monophosphate (AMP). In an intact cell, at a pH of about 7, the ATP molecule is highly charged; each of the three phosphate groups is ionized and the molecule, therefore, has four negative charges. The molecule can easily form complexes with divalent cations such as Mg^{2+} and Ca^{2+}. The result of this is that living cells have little ATP present as a free anion; most of it is bound to Mg. This feature may have something to do with the specific enzymatic hydrolysis of ATP by which the chemical potential energy is converted into work.

Free energy of hydrolysis
The energy-carrying function of ATP is not connected to the chemical bonds

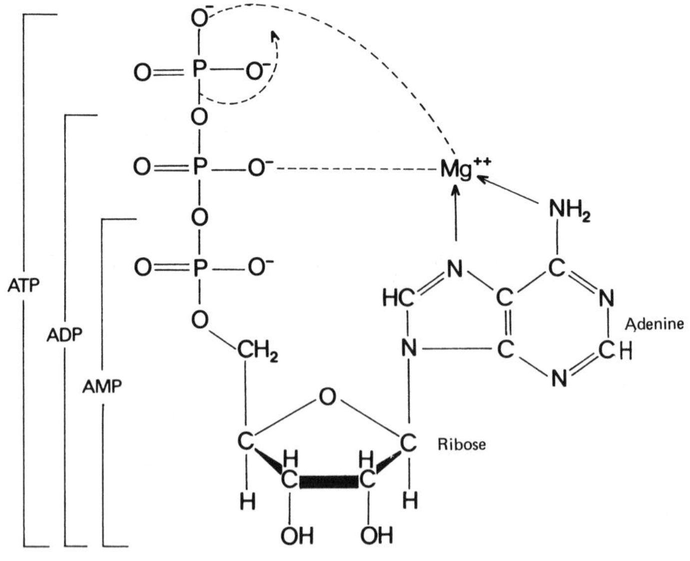

Fig. 5.3. The chemical structures of adenosine triphosphate (ATP), adenosine diphosphate (ADP), and adenosine monophosphate (AMP). ATP can easily form a complex with Mg^{2+} as shown. The molecule then is twisted and folded somewhat.

146

of the phospate groups, as may be suggested by the inaccurate term "high-energy phosphate bond". It is due to the strongly negative free energy of the hydrolysis reaction

$$ATP^{4-} + H_2O \rightarrow ADP^{3-} + HPO_4^{2-} + H^+$$

which, at equilibrium under standard condition, is about -7 kcal/mole. The equilibrium of the reaction is far to the right because the reaction products are stabilized as a result of their negative charges and the formation of new hybrid molecular orbitals. ATP is not the only phosphate compound in biological systems having this feature. In fact, there are many other phosphate compounds which have higher (and many more which have lower) standard free energies of hydrolysis. Table 5.1 summarizes the standard free energy of hydrolysis of a number of compounds in biological systems. We can see from this table that the value for ATP is actually somewhere in the middle of the range of standard free energies of hydrolysis. This makes it very suitable for the function of energy transfer by coupled reactions having common intermediates.

Coupled reactions

In coupled reactions a reaction having a negative free energy change can be used to "drive" another reaction having a positive free energy change. Consider, for example, the reactions

$$A \rightleftarrows P + Q \tag{5.6}$$

and

$$P + B \rightleftarrows R . \tag{5.7}$$

The free energy change of reaction (5.6) is

$$\Delta G_1 = - \mu_A + \mu_P + \mu_Q = - \mu_A^\circ + \mu_P^\circ + \mu_Q^\circ + RT\ln C_P C_Q / C_A \tag{5.8}$$

and that of reaction 5.7 is

Table 5.1. Standard free energy of hydrolysis of some phosphorylated compounds*

Compound	G° (kcal)
Phosphoenolpyruvate	– 14.80
1,3-Diphosphoglycerate	– 11.80
Phosphocreatine	– 10.30
Acetyl phosphate	– 10.10
Phosphoarginine	– 7.70
ATP	– 7.30
Glucose 1-phosphate	– 5.00
Fructose 6-phosphate	– 3.80
Glucose 6-phosphate	– 3.30
Glycerol 1-phosphate	– 2.20

* From A.L. Lehninger, "Biochemistry", Worth Publishers, New York, 1970, p. 302.

$$\Delta G_2 = -\mu_P - \mu_B + \mu_R = -\mu_P^\circ - \mu_B^\circ + \mu_R^\circ + RT\ln C_R/C_P C_B \quad (5.9)$$

in which the μ's are the appropriate chemical potentials and the C's designate concentrations (or rather, activities). We can assume conditions under which ΔG_1 is negative and ΔG_2 is positive. Such conditions could exist when, for instance, the concentration of P (which is common to both reactions) is very low, thus giving μ_P a large negative value. Such conditions would make reaction 5.6 go to the right and reaction 5.7 to the left until the concentrations of all reactants reached values which make both ΔG_1 and ΔG_2 zero. The free energy liberated would be dissipated in the form of heat and the only result would be an increase in the temperature of the solution. If, however, the substance P formed in reaction 5.6 is not allowed to enter the solution but is used directly in reaction 5.7 to form R, the two reactions are coupled and the total free energy change will be the algebraic sum of ΔG_1 and ΔG_2. If this sum is negative the overall reaction

$$A + B \rightleftarrows Q + R \quad (5.10)$$

proceeds to the right, thus ensuring that the free energy change of reaction 5.6 is used to do useful work (synthesizing R from B in reaction 5.7).

Many coupled reactions of this sort take place in biological systems and ATP is involved in most of them. An example is the reaction in which the energy liberated during enzymatic oxidation of 3-phosphoglyceraldehyde is stored in ATP. This reaction is a part of glycolysis, the anaerobic breakdown of sugar in cells (see section 5.3). In this reaction the aldehyde is not oxidized directly into carboxylic acid, but rather is first oxidized (in the presence of phosphate) into an intermediate called 1,3-diphosphoglycerate. If we denote the aldehyde by

$$P_i - R \overset{\displaystyle O}{\underset{\displaystyle H}{\big\backslash}}$$

(in which P_i stands for phosphate), the reaction can be represented by

$$P_i - R \overset{\displaystyle O}{\underset{\displaystyle H}{\big\backslash}} + P_i^{2-} \rightleftarrows P_i - R \overset{\displaystyle O}{\underset{\displaystyle P_i^{2-}}{\big\backslash}} + 2H . \quad (5.11)$$

The standard free energy change of this reaction is about -7 kcal/mole but since the diphosphoglycerate concentration may initially assumed to be low, the actual free energy change has a larger negative value. Reaction 5.11 is coupled to the phosphorylation of ADP by the diphosphoglycerate and the formation of phosphoglycerate and ATP:

148

$$P_i - R\overset{\displaystyle O}{\overset{\|}{-}} P_i^{2-} + ADP^{3-} \rightleftharpoons P_i - R\overset{\displaystyle O}{\overset{\|}{\underset{O^-}{\diagdown}}} + ATP^{4-} \qquad (5.12)$$

Since this reaction has a standard free energy change of some 7 kcal/mole, it can proceed to the right and the coupling of both reactions ensures that ATP is synthesized by expending the free energy change of the oxidation of the phosphoglyceraldehyde. The synthesis of sucrose from glucose and fructose is a process in which the hydrolysis of ATP is coupled to a synthetic reaction. This reaction requires energy; the standard free energy is +5.5 kcal/mole. ATP provides this energy by first phosphorylating the glucose into glucose 1-phosphate which then reacts with fructose to form sucrose and inorganic phosphate. The function of the ATP/ADP couple as energy and phosphate "conveyers" is clearly demonstrated by these reactions.

The molecular mechanism of the coupling of reactions as described is largely unknown. Enzymes are required in all known cases, many of which are found in or on membranes. Enzymes that are involved in reactions coupled to the hydrolysis of ATP are called *ATPases*. Examples are $Na^+K^+ATPases$ in membranes (which mediate the active transport of the cations Na^+ and K^+ across membranes using the energy of hydrolysis of ATP), actomyosin (that is involved in locomotion) and the coupling factor in mitochondria, chloroplasts, and bacterial systems. These systems will be discussed in more detail in later chapters.

Pyridine nucleotides
As has been stated before, the formation of ATP is always coupled to oxidation reactions having a relatively large negative free energy change. These reactions are invariably enzymatic reactions and often also require the presence of coenzymes. The two pyridine nucleotides, *nicotinamide adenine dinucleotide* (NAD) and *nicotinamide adenine dinucleotide phosphate* (NADP), are among the most common coenzymes. These cofactors mediate the oxidation-reduction reactions by acting as electron carriers. The oxydation of the phosphoglyceraldehyde in glycolysis, for instance, is mediated by NAD which itself, becomes reduced in the process. The structures of NAD and NADP are given in Fig. 5.4. Both are dinucleotides, consisting of two sugar molecules (D-ribose) connected to each other by two phosphate groups. Adenine is attached to one of the riboses, which in the case of NADP contains an additional phosphate, and the other ribose holds the base nicotinamide. Oxidation-reduction reactions take place at the nicotinamide. In the oxidized form the nitrogen in the nicotinamide ring bears a positive charge. Reduction, which requires two electrons and a proton, neutralizes the charge and adds a hydrogen to the ring, Thus, the oxidation of phosphoglyceraldehyde occurs concomitantly with the reduction of NAD^+ and the overall equation becomes

$$P_i - R\overset{\displaystyle O}{\overset{\|}{\underset{H}{\diagdown}}} + P_i^{2-} + NAD^+ + ADP^{3-} \rightarrow P_i - R\overset{\displaystyle O}{\overset{\|}{\underset{O^-}{\diagdown}}} + H^+ + NADH + ATP^{4-} \qquad (5.13)$$

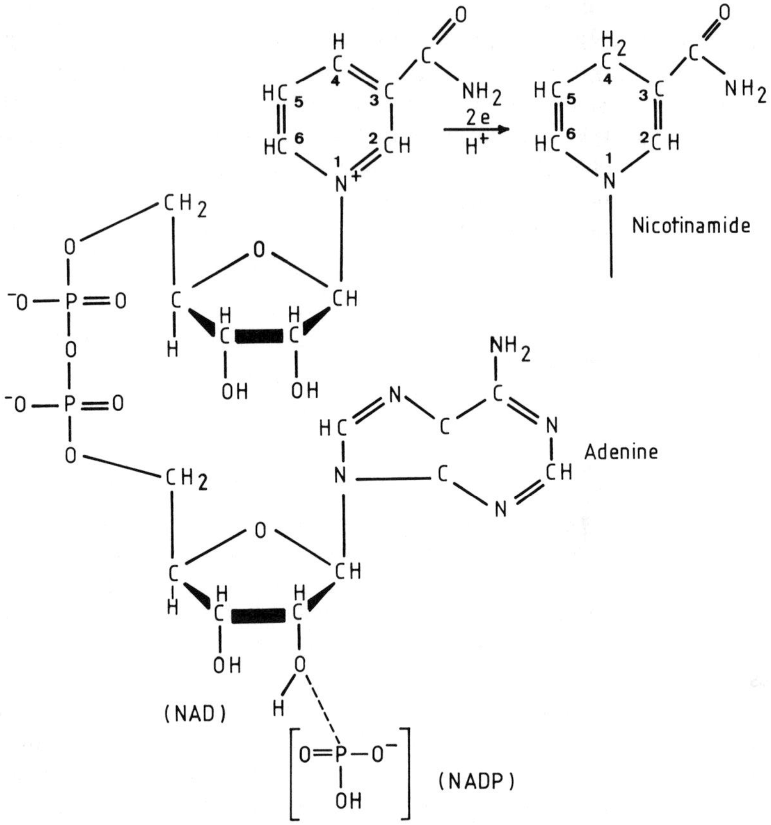

Fig. 5.4. The chemical structure of nicotinamide adenine dinucleotide (NAD) and nicotinamide adenine dinucleotide phosphate (NADP). In NADP the hydroxyl on the 2nd C atom in the ribose of the adenine nucleotide is replaced by a phosphate (structure in brackets). In the oxidized form the dinucleotides carry a positive charge on the nitrogen in the 1-position of the nicotinamide. Two electrons and a proton (in fact a hydride ion, H^-) from a reducing substrate cause the reduction of the 1 and the 4 positions of the base.

or, written in a way in which it is often presented,

$$P_i - R\begin{matrix} O \\ // \\ \\ \backslash \\ H \end{matrix} \xrightarrow[NAD^+ \;\curvearrowright\; NADH]{ADP + P_i \;\curvearrowright\; ATP} P_i - R\begin{matrix} O \\ // \\ \\ \backslash \\ O^- \end{matrix} + H^+ . \tag{5.13a}$$

Transhydrogenase

Although the two pyridine nucleotides NAD and NADP have near structural identity and both are oxidation-reduction mediators, their biological functions seem to differ somewhat. In general, NADP is used for reductive syntheses while NAD is more functional in energy metabolism. The cell, however, has an available mechanism by which NADPH can be converted to NADH and *vice versa*. This is an enzymatic reaction and the enzymes are called *transhydro-hydrogenases*. Measurements of the levels of NADPH and NADH in intact mitochondria have shown that under conditions in which energy is available the level of NADPH is in excess of that of NADH. This suggests that hydrogen is transferred from NADH to NADPH when it is not needed for further conservation of energy and can be used for reductive syntheses. The cell thus would have means of coordinating the synthesizing and energy delivering processes.

5.3 Fermentation and glycolysis

Anaerobic oxidation
In biological systems, energy is always recovered in the form of ATP, the synthesis of which is coupled to a stepwise oxidation of food. Such an oxidation can occur even in the complete absence of oxygen. In some organisms such an anaerobic oxidation process is the only form of energy conversion. In most cells, however, anaerobic as well as aerobic energy conversion can take place and in the cells of all higher organisms an obligatory anaerobic step precedes the aerobic step.

The anaerobic oxidation of sugar (or amino acids or fatty acids) is called *fermentation*. There are several forms of fermentation, with different initial substrates and different final products. Some yeast cells live anaerobically by fermenting glucose into ethanol. Some bacteria can ferment glucose into acetone or butanol, while other bacteria ferment glucose into lactic acid. The latter process is the most widespread and best understood. It is called *glycolysis* and is the type of glucose breakdown that precedes the further sequence of oxidations in higher organisms.

Glycolysis
Glycolysis occurs in the cytoplasm of the cell. It is a series of enzymatic reactions in which the six-carbon glucose is phosphorylated, isomerized, and again phosphorylated after which it breaks down into two three-carbon fragments which are readily interconverted. One of these fragments, 3-phosphoglyceraldehyde, is then oxidized with the formation of ATP, and after another isomerization and an elimination of a water molecule another phosphorylation step occurs which is coupled to an intramolecular oxida-tion-reduction reaction and the formation of ATP. Figure 5.5 give the details of this sequence. Glucose is phosphorylated into glucose 6-phosphate by an enzyme called hexokinase. This phosphorylation goes on at the expense of the third phosphate of an ATP molecule. After the isomerization into fructose 6-phosphate (by the enzyme phosphoglucomutase) another phosphorylation, at the expense of another ATP molecule, takes place. The enzyme catalyzing this process is phosphofructokinase. The product at this stage is fructose

151

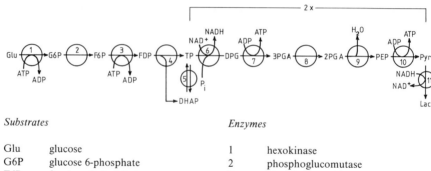

Fig. 5.5. The glycolitic sequence.

diphosphate which can now be split, by the enzyme aldolase, into the two triose phosphates, 3-phosphoglyceraldehyde and dihydroxyacetone phosphate. The process, thus, needs to be "primed"; in order to be able to synthesize ATP it must use ATP. Of course, more ATP has to be produced than expended, otherwise the situation would not make any sense. In fact, if we follow the sequence further, we see that the 3-phosphoglyceraldehyde is oxidized by NAD^+ into 3-phosphoglycerate, *after* first being phosphorylated, and that the phosphorylated oxidized 1,3-diphosphoglycerate in turn phosphorylates ADP into ATP. The enzyme catalyzing this step is glyceraldehyde 3-phosphate dehydrogenase (this reaction was used as an example of coupled reactions in section 5.2). The 3-phosphoglycerate is then isomerized (the enzyme is phosphoglyceromutase) and converted into phosphoenolpyruvate (with enolase) eliminating a water molecule. Finally, phosphoenolpyruvate phosphorylates another ADP molecule into ATP (with pyruvate phosphokinase) producing pyruvate. Two molecules of ATP, thus, are synthesized in one sequence from 3-phosphoglyceraldehyde to pyruvate. But since the enzyme triose phosphate isomerase converts the other triose phosphate into 3-phosphoglyceraldehyde, four molecules of ATP are synthesized and two molecules of ATP are expended for each sequence of glucose-to-pyruvate. The net yield, thus, is two molecules of ATP per glucose molecule.

The aldolase reaction, which results in the cleavage of fructose diphosphate, has an equilibrium of about 90% in the direction of the fructose diphosphate. The reason why the reaction goes forward is that of one of its reaction products,

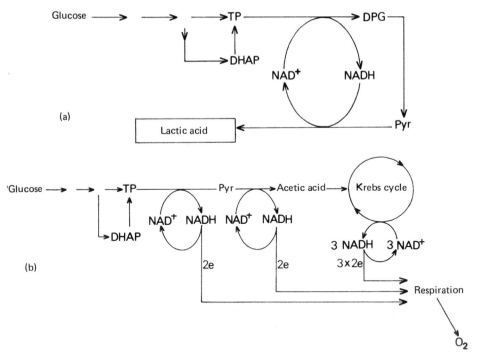

Fig. 5.6. (*a* The cyclic nature of the NAD reaction in anaerobic metabolism. (*b*) The electron intermediate function of NAD in aerobic metabolism.

3-phosphoglyceraldehyde, is removed by oxidation, the energy of which is conserved in the production of ATP. Thus, for the glycolytic sequence to proceed at all it is necessary that the 3-phosphoglyceraldehyde be oxidized. As we will see later, this provides a means of control.

When glycolysis is not followed by aerobic energy conversion (in glycolytic anaerobic cells or in facultative cells in the absence of oxygen) the pyruvate is reduced into lactate; in this reaction NADH is oxidized. Thus, in anaerobic metabolism NAD^+ reacts in a cyclic manner (see Fig. 5.6a) (there is no net oxidation, but energy is conserved from an oxidative process all the same), while in aerobic metabolism NAD^+ is further reduced. (Fig. 5.6b).

5.4 The citric acid cycle

Aerobic oxidation
In aerobic cells glycolysis is followed by the respiratory reactions which occur in two groups of steps. The first step is a cyclic series of enzymatic reactions. This cycle is the *citric acid cycle* (also known as Krebs cycle after its discoverer, Hans Krebs). A number of decarboxylations and oxidations take place in the citric acid cycle. The reduced products of these reactions (NADH and

153

reduced flavin) enter the second set of reactions, the *respiratory chain*, which is bound to cristae. The final oxidation to the level of oxygen and the membrane-coupled synthesis of ATP occur through this chain.

In eukaryotic cells, these two sets of reactions occur inside the mitochondria (Fig. 5.7). This organelle consists of two membranes: the outer membrane is permeable to most smaller molecules and the inner membrane convolutes to the inside forming the many inward folds called cristae (see also Fig. 2.6). The surface of this membrane is thus tremendously increased. It encloses an inner compartment, the matrix, and is permeable only to water and a limited number of small neutral molecules, such as urea an glycerol. The membrane contains, however, several permeases, "enzymes" which are carriers for specific metabolites such as amino acids, acetates, NADPH, NADH, ADP, ATP, phosphates, and others.

The cycle
Although the principal role of the citric acid cycle is the coupling of the glycolytic breakdown of sugar in the cell cytoplasm to the series of oxidation-reduction reactions leading to the production of ATP in the mitochondria, it also serves to regulate the synthesis of a number of compounds required by the cell. Several intermediates of the cycle are branch points from which the synthesis of, for instance, amino acids and fat can start. Many of these "branching off" reactions are reversible, so that they can also serve to generate the cycle intermediates.

In the citric acid cycle the 4-carbon compound oxaloacetate incorporates a 2-carbon acetyl group to become the 6-carbon citric acid. Subsequent decarboxylations and oxidations then lead to the regeneration of oxaloacetate, thus closing the cycle. The acetyl group is furnished by glycolysis. To this end the pyruvate is not reduced to lactate, as in anaerobic cells, but oxidized and decarboxylated (CO_2 is liberated) by the enzyme pyruvate dehydrogenase (see Fig. 5.6b). NAD^+ is again a coenzyme and the reaction is, thus, coupled to its reduction. The oxidation product is not free acetate but an acetyl group which is fused to another coenzyme called *coenzyme* A (CoA). Coenzyme A

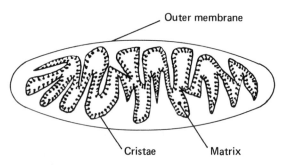

Fig. 5.7. A representation of a cross section of a mitochondrion. The inner membrane convolutes to the inside, the matrix, thus forming the many folds called cristae. The components of the respiratory chain are embedded in the cristae.

154

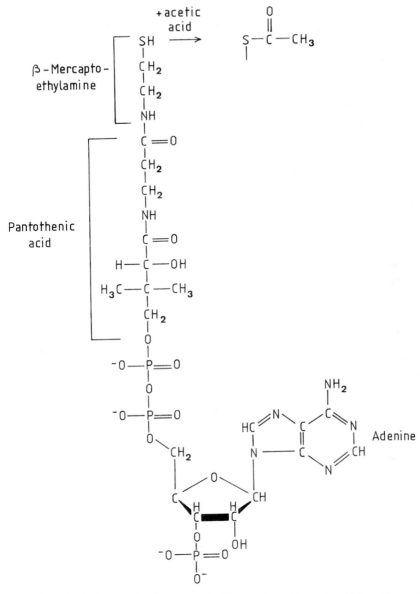

Fig. 5.8. The chemical structure of coenzyme A. The reaction with acetic acid (or with succinic acid) takes place at the SH terminal of the molecule, thus forming acetyl-S-CoA (or succinyl S-CoA).

is a complex molecule consisting of a base (adenine), a sugar (ribose), phosphate groups, and a tail to which a sulfhydryl group is attached (see Fig. 5.8). The sulfhydryl is the active part and the molecule is usually written as CoASH; the reaction with pyruvate (from glycolysis) and NAD^+, thus, produces acetyl-S-CoA.

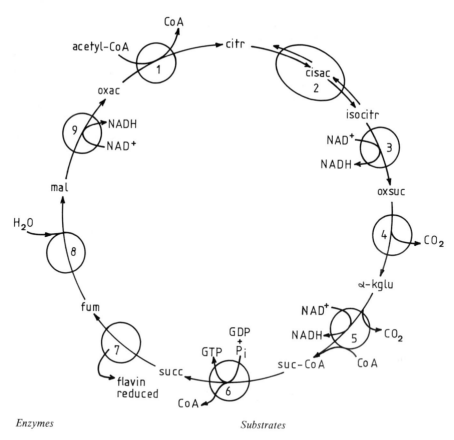

Enzymes

		Substrates	
citr	citric acid	1	citrate synthase (condensing enzyme)
cisac	*cis*-aconic acid	2	aconitase
isocitr	isocitric acid	3	isocitrate dehydrogenase
oxsuc	oxalosuccinic acid	4	oxalosuccinate decarboxylase
α-kglu	α-ketoglutarate	5	α-ketoglutarate dehydrogenase
suc-CoA	succinyl coenzyme A	6	succinyl-CoA synthase
succ	succinic acid	7	succinate dehydrogenase
fum	fumaric acid	8	fumarase
mal	malic acid	9	malate dehydrogenase
oxac	oxaloacetic acid		

Fig. 5.9. The citric acid cycle.

Acetyl-S-CoA then enters the cycle by reacting with oxaloacetate to form citrate and free CoA-SH (Fig. 5.9). The first oxidation of the cycle is preceded by an isomerization of the citrate into isocitrate. Then the isocitrate is oxidized into oxalosuccinate concommitant with the reduction of NAD^+. The next step is a decarboxylation; CO_2 is again removed and α-ketoglutarate is formed. This compound is the first branch point of the cycle; α-ketoglutarate is a precursor for a number of amino acids in reactions called *transamination reactions*. In the cycle itself the α-ketoglutarate is decarboxylated and oxidized

156

in one reaction. NADH is again the reduced product and the succinyl group which is formed is attached to CoA-SH in exactly the same manner as the acetyl group. The succinyl-S-CoA, thus formed, is another branch point of the cycle; from it proceeds the synthesis of porphyrin, the backbone of heme, which is the prosthetic group of cytochromes and hemoglobin. Succinyl-S-CoA, as well as acetyl-S-CoA, are intermediates in fatty acid metabolism. The cycle proceeds with the formation of succinate. This reaction is an energy-conserving step; it is coupled to the phosphorylation of guanosine diphosphate (GDP) into guanosine triphosphate (GTP). The GTP then undergoes a phosphate group transfer reaction with ADP:

$$GTP + ADP \rightleftarrows GDP + ATP . \qquad (5.14)$$

The next step of the cycle is the oxidation of succinate into fumarate. This reaction differs from the other oxidation-reduction (dehydrogenase) reactions of the cycle; its reaction partner is not the coenzyme NAD but rather the succinate dehydrogenase enzyme itself. Succinate dehydrogenase is a protein with flavin adenine dinucleotide (FAD) as a prosthetic group. When succinate is oxidized, FAD is reduced to $FADH_2$. This enzyme is tightly bound to the mitochondrial membranes and is an entrance port into the respiratory chain for electrons, as we will see later. Fumarate is involved in nitrogen metabolism. The enzyme fumarase then takes care of the hydration of fumarate into malate and a final oxidation step (again concomitant with the reduction of NAD^+) regenerates oxaloacetate, ready to react again with another acetyl-S-CoA molecule. Figure 5.9 gives a summary of the reaction cycle, including the names of the enzymes involved.

Anaplerotic reactions

The pivotal function of the citric acid cycle, in the first place for the aerobic oxidation of sugar and in the second place for provision of intermediates for biosynthesis, makes it necessary that the cycle be kept running within close tolerances. First, the siphoning off of intermediates for biosynthesis must be compensated for lest the cycle come to a halt. Part of such a compensation is accomplished by the fact that many of the branching-off reactions are reversible, so that cycle intermediates can be synthesized from amino acids via these reversed pathways. But there are other special, so-called *anaplerotic* ("filling up") reactions. The most important of these is the enzymatic carboxylation of pyruvate into oxaloacetate:

$$Pyruvate + CO_2 + ATP \rightarrow oxaloacetate + ADP + P_i . \qquad (5.15)$$

The enzyme catalyzing the reaction is *pyruvate carboxylase* and is primarily found in the mitochondria of the liver cells of most species. This reaction is a very good example of the use of the allosteric properties of enzymes for control. Pyruvate carboxylase is an allosteric enzyme with acetyl-CoA as a positive modulator; thus, the higher the level of acetyl-CoA, which is the fuel for the citric acid cycle, the better the reaction rate, the more oxaloacatate produced, and the more acetyl-CoA oxidized in the cycle. In the absence of acetyl-CoA, the rate of reaction is very low.

Another anaplerotic reaction is the synthesis of malate from pyruvate by the malic enzyme, a reaction which involves the oxidation of NADPH. Many plants and bacteria can also use a sort of short circuit of the citric acid cycle to produce more succinate and malate directly from isocitrate. In this so-called *glyoxylate cycle*, isocitrate is split into succinate and glyoxylate by the enzyme *isocitritase*. Another enzyme, *malate synthase*, then promotes the reaction of glyoxylate wich acetyl-CoA to form malate and CoA-SH.

5.5 Respiration

The respiratory chain

Most of the electrons involved in the oxidation reactions of the citric acid cycle end up in NADH. Only those from the reaction of succinate are in $FADH_2$ which is bound to the succinate dehydrogenase enzyme. NADH is also a product of the dehydrogenation of pyruvate. Moreover, for every glucose molecule entering the glycolytic chain, two NADH molecules are produced in the dehydrogenation step. All these NADH molecules react with the enzyme NADH dehydrogenase. NADH dehydrogenase, like succinate dehydrogenase, is a flavin-linked enzyme; its prosthetic group is flavin mononucleotide (FMN) which is reduced when NADH is oxidized. Both hydrogenases are bound to the membrane and can be seen as "entrance ports" to the respiratory chain.

The respiratory chain is a series of oxidation-reduction enzymes and coenzymes which actually take part in the reactions themselves. The chain

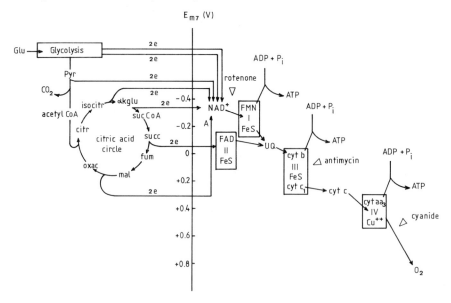

Fig. 5.10. The respiratory chain.

158

consists, in addition to the two dehydrogenases mentioned above, of a number of polypeptides and cofactors, some integral protein complexes of the cristae membrane and some peripherally bound to it. Among these are a coenzyme called *ubiquinone, iron-sulfur proteins* and *cytochromes*. Iron-sulfur proteins are proteins in which the iron is covalently bound to the sulfur-containing amino acids of the peptide chains, often forming clusters of two or four groups called *iron-sulfur centers*. Cytochromes are proteins with a *heme* as prosthetic group. A heme group is a tetrapyrole with iron chelated in its center (Fig. 5.11). An oxidation (reduction) of a cytochrome (and also of an iron-sulfur center) is the removal (addition) of an electron from (to) the iron which, thus, makes a ferrous-ferric (ferric-ferrous) transition.

Traditionally, the chain was thought to be arranged as a linear sequence of components in the order of increasing redox potential. The electrons, coming from NADH and succinate were seen as moving down the redox potential gradient to oxygen, as shown by the arrows in the right-hand part of Fig. 5.10. The reduced oxygen then forms water with protons from the matrix. The situation turned out to be more complex, however. When the cristae of mitochondria are gently dissected, the membrane dissociates into five macromolecular complexes of reproducible composition. One of these complexes is the so-called *coupling factor* which is involved in the synthesis of ATP. It will be discussed below and in Chapter 8. The other four complexes contain the electron transport components. In Fig. 5.10 they are numbered I-IV. Complex I contains NADH dehydrogenase (FMN) and a number of distinct iron-sulfur centers. Complex II is a complex of succinate dehydrogenase

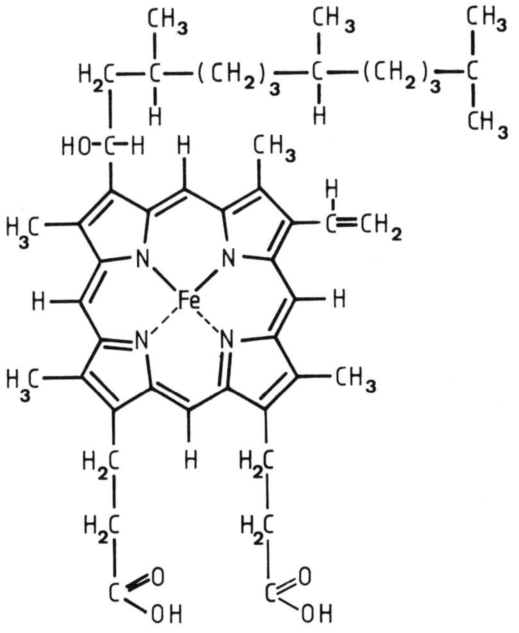

Fig. 5.11. The chemical structure of the heme group of cytochrome *c*.

(FAD) and, again a number of iron-sulfur centers. Complex III, also called *Cytochrome b/c₁ complex*, contains, probably two, cytochrome *b* species, cytochrome c_1 and again an iron-sulfur center. Complex IV contains the cytochrome aa_3 complex and copper. The membrane contains, furthermore, two components that are easily extracted, *ubiquinone* and *cytochrome c*. Ubiquinone resides within the membrane in which it can move in a lateral direction. Presumably, cytochrome *c* is peripherally bound at the cristae membrane. It has been purified and crystallized and is, consequently, the best characterized cytochrome so far.

Ubiquinone and cytochrome *c* function like shuttles. As indicated in Fig. 5.10, ubiquinone transports electrons from the complexes I and II to complex III and cytochrome *c* carries them from complex III to complex IV. The latter complex, therefore, is often called *cytochrome c oxidase*. The mechanisms of electron transport within the complexes are not known and also the role of ubiquinone may be more complex than that of a simple linear electron transporter (see Chapter 8).

Much of what is known about repiratory redox reactions could be deduced from applications of sensitive absorption difference spectrophotometry. Cytochromes absorb light in the visible spectral region and their spectra undergo substantial changes when the enzyme changes its redox state. Figure 5.12 shows the spectrum of cytochrome *c* in the reduced and oxidized states. Using

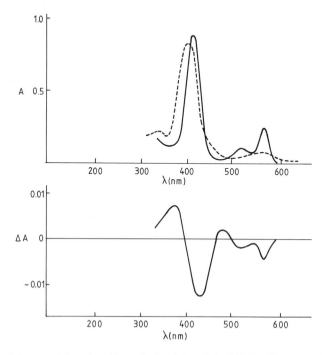

Fig. 5.12. (*top*) Spectra of the reduce (drawn line) and the oxidized (dashed line) form of cytochrome *c*; (*bottom*) oxidized minus-reduced difference spectrum of cytochrome *c*.

160

absorption difference spectrophotometry one can follow the oxidation-reduction reactions of the cytochromes and, thus, determine the sequence of the oxidation-reduction reactions of the electron carriers. A difference spectrum of respiring versus nonrespiring rat liver mitochondria is given in Fig. 5.13. These processes can be more closely examined by using specific inhibitors. The insecticide *rotenone*, for example, specifically blocks the electron transport

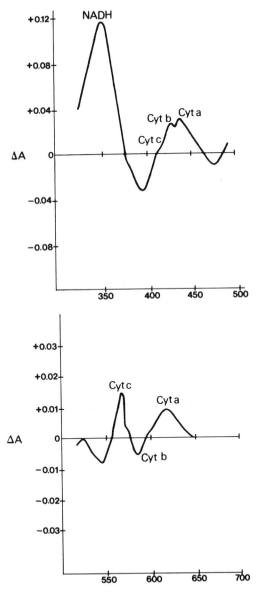

Fig. 5.13. Difference spectrum of respiring-minus-nonrespiring rat liver mitochondria (from A.L. Lehninger, Biochemistry," Worth Publishers, New York, 1970, p. 380).

from NADH to flavoprotein. Other inhibitors are the antibiotic *antimycin*, which blocks between the cytochromes *b* and *c* and *cyanide*, which prevents oxygen from oxidizing cytochrome *c* oxidase. Using such blocks it is easy to demonstrate that all components before the block in the direction of oxygen are completely reduced while all components after the block remain oxidized.

Oxidative phosphorylation

The negative change in the free energy of the redox reactions in the chain is conserved in the form of ATP. We can distinguish three regions in the chain in which the difference in redox potential between consecutive components is relatively large. Thus, the difference in redox potential between NADH and ubiquinone is some 0.27 V, that between cytochrome *b* and cytochrome *c* is about 0.22 V, and that between the cytochrome aa_3 complex and oxygen is 0.53 V. Each of these three energy gaps is sufficiently large to provide for the free energy of the ADP→ATP phosphorylation reaction. This would suggest three sites at which the phosphorylation could be coupled to electron transport. In fact, it has been demonstrated that for each couple of electrons from NADH, three molecules of ATP are formed and that a couple of electrons from succinate can provide for only two ATP molecules. This, then, would be consistent with three "coupling sites", as indicated in Fig. 5.10: one in the oxidation of NADH, the second between ubiquinone and cytochrome *c* and the third associated with cytochrome aa_3 complex. Originally, the redox reactions at these very sites were thought to be coupled to the phosphorylation reaction; the respiratory reactions and phosphorylation were thought to be *chemically* linked. The now generally accepted *chemiosmotic hypothesis*, proposed in the 1960's by P. Mitchell, departs from a different, more *physical* concept: according to this hypothesis, the coupling between the redox reactions in the respiratory chain and the phosphorylation of ADP to ATP is by a trans-membrane current of protons. In this model, which will be discussed in detail in Chapter 8, the "coupling sites" located in three of the four respiratory complexes are sectors of the respiratory chain in which a "proton pump" is active. The fifth integral membrane protein complex, the coupling factor, is an ATPase or rather an ATPsynthase in which a reversed proton current causes the synthesis of ATP.

The coupling factor

The fifth protein complex that can be isolated from the mitochondrial inner membrane is the coupling factor. The search for it was particularly motivated by the need to find a so-called *high-energy intermediate*. The existence of such an intermediate was suggested by the peculiar action of some inhibitors of the phosphorylation process. The coupling between respiratory electron transport and phosphorylation can be broken by a number of compounds that either inhibit the whole process or only stop the synthesis of ATP while leaving the electron transport going, often even stimulating it. Compounds exhibiting the latter effect are called *uncouplers*. An effective uncoupler for mitochondrial phosphorylation is 2,4-dinitrophenol (DNP). A certain class

162

of antibiotics are also uncouplers of phosphorylation in both respiratory and photosynthetic systems. The uncoupling effect of such compounds could not be understood without accepting the synthesis or generation of a "high-energy intermediate" by electron transport preceding the phosphorylation of ADP to ATP. At present, it is generally accepted that the high-energy intermediate is not a chemical entity but a membrane condition (in fact, an electrochemical potential created by the proton current; see Chapter 8). Uncouplers act by promoting proton leaks in the membrane, thus preventing a reversed proton current through the ATPsynthase.

The isolated coupling factor nevertheless appeared essential: membrane fractions devoid of the factor have lost their ability to synthesize ATP but still carry out electron transport. When purified coupling factor is added to the deficient fractions, ATP synthesis is, at least partially, restored. It turned out that the coupling factor is the enzyme complex that converts the electrochemical potential energy across the membrane into the chemical potential energy of ATP (see also Chapter 8).

Presently, the complex has been isolated from mitochondria as well as from chloroplasts and some photosynthetic prokaryotes. Figure 5.14 is a electron micrograph of a fraction of mitochondrial cristae, made by E. Racker. It shows the complex as little spherical particles with a diameter of 80–90 Å, sticking out toward the inside of the matrix space and connected to the inner membrane by a narrow stalk. The stalks sitting in the membrane are designated

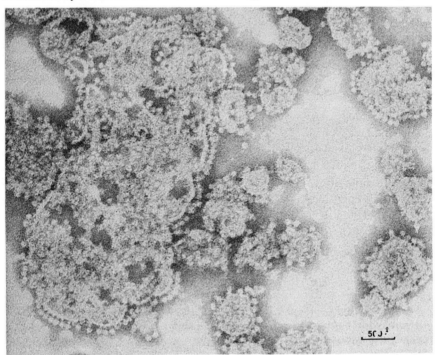

Fig. 5.14. Electronmicrograph of a preparation of a mitochondrion showing the coupling factor as little knobs on talks sticking out into the matrix (courtesy of Dr. E. Racker, Cornell University).

as F_0 and the spherical parts sticking out as F_1 (CF$_1$ for chloroplasts). The complete complex often is called F_0F_1 ATPase or F_0F_1 ATP-synthase. Although in some cases its structure is reasonably well known (see Chapter 8) the mechanism(s) by which it functions is still elusive.

Control mechanisms

Fermentation, in particular glycolysis, and respiration in living cells must proceed at rates and levels which are adapted to specific functions and conditions. This is most obvious for the eukaryotic cells of the higher organisms; thus, in liver cells, for example, these requirements are very different from those of muscle cells. Different conditions set different requirements in prokaryotic cells as well. To provide for such flexibility, both enzyme systems must operate under tight controls. Control of glycolysis is already affected by the rate at which the product of the aldolase reaction, 3-phosphoglyceraldehyde, is removed by the oxidation step. This is dependent on the level of the NAD$^+$ which, under anaerobic conditions, is regenerated from NADH by the reduction of pyruvate to lactate. Thus, in facultative cells, utilizing glucose from an ample supply under anaerobic conditions, glycolysis runs at a high rate and lactate is accumulated. But when such cells are transferred into aerobiosis the rate of glycolysis decreases dramatically and the accumulation of lactate is reduced to zero. This inhibition of glycolysis by oxygen is called the *Pasteur effect* after Louis Pasteur, who discovered it during his investigations of the fermentation processes of wine making. Since respiration is reoxidizing NADH, thus leading to an increased NAD$^+$/NADH ratio, one would expect a stimulation rather than an inhibition of glycolysis by oxygen. Of course, a controlling effect is exerted by the "shuttle" mechanism that forms the communication link between the cytoplasmic pool and the intramitochondrial (matrix) pool of NAD$^+$ and NADH. However, a major factor of control is that the activity of one of the key enzymes of glycolysis, phosphofructokinase, is allosterically inhibited by ATP and stimulated by ADP and phosphate. Phosphofructokinase catalyzes the rate-limiting reaction in which fructose 6-phosphate is phosphorylated into fructose 1,6-diphosphate; when during respiration the ATP/ADP ratio increases the phosphofructokinase is gradually "turned off", thus slowing down the rate of glycolysis. Although phosphofructokinase seems to be the major regulating enzyme of glycolysis, there are a number of secondary control points in the glycolytic sequence which are influenced by the ATP/ADP ratio as well as by the level of intermediates such as citrate, acetyl-CoA and glucose 6-phosphate.

The ATP/ADP ratio not only controls the rate of glycolysis but also has a profound influence on the rate of respiration. When the supply of respiratory substrates is ample, a high rate of oxygen consumption occurs when the ADP and phosphate concentration is high and the concentration of ATP is low. When the concentration of ATP rises and the concentrations of ADP and phosphate are reduced to zero the respiratory rate becomes very low. This effect is called *respiratory control* or *acceptor control*. This can easily be

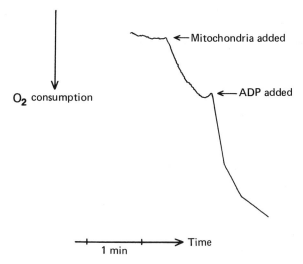

O_2 consumption

←Mitochondria added

←ADP added

1 min → Time

Fig. 5.15. An experiment showing respiratory control. The uptake of oxygen increases sharply when ADP is added to a suspension of respiring mitochondria.

demonstrated by an experiment such as that schematically diagrammed in Fig. 5.15. Mitochondria, in the presence of phosphate but in the absence of ADP, show a very low rate of oxygen consumption. When ADP is added the rate increases by as much as a factor of 20. When all added ADP is phosphorylated the rate returns to the original low level.

The chemiosmotic model offers a simple explanation for respiratory control (see also Chapter 8). If phosphorylation cannot proceed by lack of the substrates ADP and/or phosphate, the electrochemical potential created by the proton current rises until it approaches the value of the free energy of the oxidation; the rate of respiration then slows down to a rate that is determined solely by leakage of protons through the membrane. As soon as phosphorylation resumes upon addition of the substrates the electrochemical potential is relieved by the reversed proton current through the ATPsynthase, and respiration accelerates. The fact that uncouplers of phosphorylation such as dinitrophenol, stimulate the rate of respiration is supporting evidence for this explanation. Uncouplers are supposed to cause a breakdown of the electrochemical proton potential by causing proton leaks in the membrane.

The relative concentrations of ATP and ADP in the cell are, thus, the most important controlling elements. This is true for all processes generating or utilizing the energy incorporated in the phosphate bond. Many regulatory enzymes, as well as those involved in biosynthetic pathways, are responsive to the levels of ATP and ADP (and also of AMP). Regulation, therefore, is accomplished in all these reactions by a delicate balance of the concentrations of these important nucleotides.

Bibliography

Ernster, L. and Schatz, G. (1981). Mitochondria: A Historical Review., *J. Cell. Biol.* **91**, 227S–255S.

Hackenbrock, C.R. (1981). Lateral Diffusion and Electron Transfer in the Mitochondrial Inner Membrane, *TIBS Rev.,* 151–154.

Krebs, H., and Kornberg, H.L. (1957). "Energy Transformations in Living Matter." Springer Verlag, Berlin.

Mitchell, P. (1961). Coupling of Phosphorylation to Electron and Hydrogen Transfer by a Chemiosmotic Type of Mechanism, *Nature* **191**, 144–148.

Racker, E. (1976). "A New Look at Mechanisms in Bioenergetics." Acad. Press, New York.

Van Dam, K., and Westerhoff, H.V. (1980). A Description of Biological Energy Transduction by "Mechanistic Thermodynamics". *Recl. Trav. Chim. Pays Bas* **99**, 329–332.

6. Photosynthesis

6.1 Photosynthetic structures

Chloroplasts, thylakoids, and chromatophores

Photosynthesis is the process by which the energy of sunlight is captured for use in sustaining life. The organisms which are capable of performing this process are plants and algae (multi, as well as unicellular), and certain kinds of bacteria. In eukaryotic photosynthetic organisms (all plants and algae) the process occurs within an organelle called the *chloroplast* (see Chapter 2). The photosynthetic apparatus is embedded in lamellae which are densely packed in some regions (grana) and in other areas extend into the stroma (intergranal lamellae). The lamellae are closed double membranes forming the flattened closed vesicles called *thylakoids* (Fig. 6.1). The photosynthetic apparatus of the photosynthetic prokaryotes (photosynthetic bacteria and cyanobacteria) is also embedded in membranous structures which in these organisms extend throughout the cell. When photosynthetic bacterial cells are ruptured one often finds membrane fragments which are formed into closed vesicles from 300 to 500 Å in diameter. These vesicles are called *chromatophores* or *chromatophore fractions* (Fig. 6.2). They retain most of the photosynthetic activity. Functional membrane fractions from cyanobacteria are more difficult to obtain. The photosynthetic apparatus in these organisms is in membranes that are folded in a more or less complicated way inside the cell. Usually these membranes are also called thylakoids.

The photosynthetic unit

In photosynthesis light is absorbed by pigment molecules (chlorophylls, carotenoids, and phycobilins, see below) but is not utilized for energy conversion immediately upon absorption. The apparatus is organized in so-called *Photosynthetic Units*, ensembles of pigment molecules within which excitation energy is transferred until it is trapped somewhere within the unit

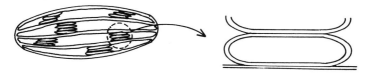

Fig. 6.1. A schematic representation of a chloroplast, showing the grana consisting of the flattened vesicles called thylakoids.

167

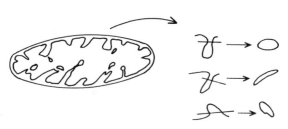

Fig. 6.2. A schematic representation of a photosynthetic bacterium showing the inward folded cytoplasmic membrane containing the photosynthetic apparatus which, upon disruption, form the closed vesicles called chromatophores.

at a site specialized for photochemical conversion. Such a site is called a *Reaction Center* (RC). The major part of the pigment molecules just serve as "harvesters" of electromagnetic energy. These pigments often are called *antenna pigments*. They are organized in large complexes with proteins.

The functioning of photosynthetic units causes a more efficient utilization of the absorbed energy; the unit can be seen as a kind of funnel feeding energy into the traps which then can turn over much faster than they would if each absorbing molecule was its own trap. In higher plants, the size of a photosynthetic unit is somewhere between 200 and 400 chlorophyll molecules per reaction center, in photosynthetic bacteria the size is about 50 bacteriochlorophyll molecules per reaction center. One could think of a number of organizations of the units. A large aggregate of pigments in which the reaction centers are scattered randomly would lead to a relatively simple description, using the Stern-Volmer equation, of the excitation energy transfer and trapping (see section 6.2). Such a model of the photosynthetic unit is often referred to as the 'lake model'. If the aggregate of antenna pigments and reaction centers are, indeed, organized in separate units, the situation becomes more complicated: in general, the rates of transfer within a unit would be different from the rates of transfer between the units. This leads to the "puddle model". It appeared that in reality there are mixtures of both models, which requires a more complex treatment of the excitation energy transfer problem.

Chlorophyll

Absorbing pigments are found in a wide variety in photosynthetic organisms. The most important of these are the chlorophylls, a group of highly conjugated structures, which differ among each other in only minor aspects. The conjugated structure is a closed tetrapyrole, called *porphyrin*, enclosing a magnesium atom. A long hydrocarbon chain called the *phytol* tail is attached to ring IV. Figure 6.3a shows the structure of chlorophyll *a*, the major chlorophyll found in all plants and algae. Figure 6.3b shows the absorption spectrum of chlorophyll *a* dissolved in ether. The major chlorophyll in photosynthetic bactaria is called bacteriochlorophyll.

168

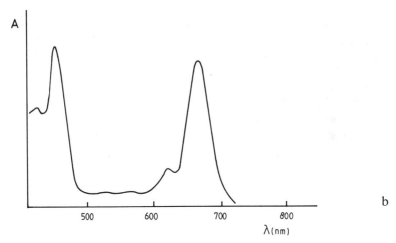

Fig. 6.3. (*a*) The chemical structure of chlorophyll *a*; (*b*) an absorption spectrum of chlorophyll *a* in organic solvent.

Accessory pigments

Most photosynthetic organisms also contain so-called *accessory* pigments. Green plants and green algae have chlorophyll *b*, red algae and cyanobacteria have phycobilins (open tetrapyroles), and all photosynthetic organisms have one or more variations of the group of carotenoids (unsaturated hydrocarbon chains with aromatic end groups at both sides). The absorption spectrum of a unicellular alga, *Chlorella pyrenoidosa* is given in Fig. 6.4. The spectrum is more or less representative, also for higher plants. The peak at 675 nm

169

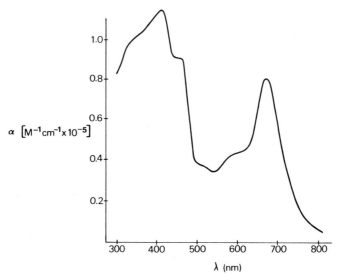

Fig. 6.4. An absorption spectrum of the green alga *Chlorella pyrenoidosa*. The maximum at 675 nm and the shoulder at about 650 nm are due to chlorophyll *a* and chlorophyll *b*, respectively.

is the *in vivo* absorption maximum of chlorophyll *a*. The shoulder at 650 nm in the spectrum is due to chlorophyll *b*; the absorption bands in the blue-green region are largely due to carotenoids. Comparison with the spectrum of chlorophyll *a* dissolved in ether (Figure 6.3b) shows that the *in vivo* absorption maximum is shifted toward the red end. This is due to the fact that the chlorophyll molecules form complexes with proteins and with each other. This complex formation results in even more profound spectral changes in some photosynthetic bacteria. A summary of the different kinds of pigments in photosynthetic organisms and their absorption characteristics *in vivo* and *in vitro* is given in Table 6.1.

6.2 Transfer and trapping of excitation energy

Sensitized fluorescence
Transfer of excitation energy is possible because of the forces involved in the redistribution of electric charge in an electronic transition; the electric dipole field of one molecule of an ensemble induces excitation in another. We can see this as a coupled event in which the deexcitation of a sensitizer molecule S is accompanied by the excitation of a different acceptor molecule A. If the latter is fluorescent, its fluorescence can be induced by exciting the sensitizer molecule:

170

Table 6.1 Photosynthetic pigments.

| Pigment | Absorption maxima (nm) | | Occurrence |
	In organic solvents	In vivo	
Chlorophyll *a*	420; 660	435; 670-680	Higher plants, algae, cyano-bacteria
Chlorophyll *b*	453-643	480; 650	Higher plants, green algae
Chlorophyll *c*	445-625	645	Diatoms, brown algae
Chlorophyll *d*	450-690	740	Some red algae
Bacteriochlorophyll *a*	365; 605; 770	585; 800-890	Purple bacteria
Bacteriochlorophyll *b*	368; 582; 795	1017	Some purple bacteria, instead of bacterio-chlorophyll *a*
Bacteriochlorophyll *c*	425; 650	750	Green bacteria
Bacteriochlorophyll *d*	432; 660	760	Some green bacteria instead of bacterio-chlorophyll c
Phycoerythrin	490; 546; 576	Same	Red algae (small amounts in cyanobacteria
Phycocyanin	618	Same	Cyanobacteria (small amounts in red algae
Carotenoids	420-525	Same	All photosynthetic organisms; some seven different kinds in different organism

$$S + h\nu_a \rightarrow S*$$
$$S* + A \rightarrow S + A*$$
$$A* \rightarrow A + h\nu_f$$

(6.1)

in which ν_a and ν_f are the frequencies of the absorbed and the emitted light, respectively. This process of sensitized fluorescence can be (and has been) easily demonstrated with pairs of dyes which do not have greatly overlapping absorption spectra. The mechanism can be likened to the transfer of energy by resonance in coupled pendulums and tuning forks. The process is, therefore, often called *inductive resonance transfer*. L. Duysens (1952) has demonstrated it in *in vivo* systems; when a photosynthetic organism is illuminated with light that is absorbed by carotenoids but not by chlorophyll, fluorescence of chlorophyll *a* occurs, thus demonstrating the transfer of excitation energy from carotenoid to chlorophyll *a*.

Excitation energy transfer
Resonance transfer is a molecular interaction which takes place on the quantum mechanical level. Therefore, it is formally incorrect to consider the excitation as localized in a particular molecule at any one time. One has to consider the excitation as a property of the whole ensemble. Wavefunctions describing the system are solutions of the Schrödinger equation

$$\mathbf{H}\Psi_j = E_j\Psi_j$$

[6.2]

171

in which the Hamiltonian has a term describing the interaction between the molecules. Solutions are feasible only when certain simplifications are made. One can, for example, ignore contributions due to intermolecular electron orbitals (as is done in the molecular exciton model) or retain only the electric dipole portion of the radiative interaction. One such simplification concerns vibrational interactions. Application of this leads to three distinguishable cases, depending on the magnitude of the interaction energy. Although formally incorrect, one can use "localized language" to describe these cases. In such language, one can speak of a transfer time τ_t which is the average time of residence of an excitation in any one molecule. If this τ_t is small as compared to the period of nuclear oscillations (about 3×10^{-14} s) and, hence, small compared to the period of the intermolecular (lattice) vibrations (about 3×10^{-12} s) as well, vibrational states do not come into play at all. The transfer is between identical electronic states of the interacting molecules and the rate of transfer is proportional to the interaction energy (hence to the inverse of the third power of the distance between the molecules in the case of dipole interaction). This is often called *fast transfer*. One speaks of *intermediate transfer* when the transfer time τ_t is between the period of nuclear vibration and that of lattice vibrations. The resonance then is among vibrational levels in the interacting molecules. Also in this case the rate is proportional to the interaction energy (third power of the inverse distance between the dipoles). In both cases, fast and intermediate transfer, a "localized treatment" leads to incorrect results. In the case of *slow transfer* the transfer time τ_t is large compared to the period of the lattice vibrations. In that case a localized treatment gives a reasonable approximation.

Molecular exciton model
M. Kasha and coworkers have developed the molecular exciton model to describe the fast transfer (ignoring vibrational interactions) in polymers. In this model the wave functions for the excited state of the polymer are linear combinations of all possible localized conditions in the polymer, thus leading to a splitting of the monomeric level with a multiplicity equal to the number of coupled monomers. Considers, for example, a dimer made up of monomers A_1 and A_2. Delocalized treatment involves a ground state $(A_1.A_2)$ and an excited state $(A_1.A_2)^*$ that is split into two levels. The wave functions for the two localized conditions $(A_1{}^* + A_2)$ and $(A_1 + A_2{}^*)$ are $\psi_1{}^*.\psi_2$ and $\psi_1.\psi_2{}^*$. Linear combinations of these yield

$$(1/\sqrt{2})\,(\Psi_1{}^*{\cdot}\Psi_2 \pm \Psi_1\Psi_2{}^*). \qquad (6.3)$$

This is illustrated in Figure 6.5 for a dimer whose transition dipole moments are perpendicular to the axis through the center of each dipole. In this case the allowed state is the upper antibonding one (positive sign in 6.3) where the transition dipoles are parallel to each other (in phase). The lower state is forbidden. That this must be so, can be seen from the following: since the dimensions of the dipole are small compared to the wavelength of the exciting light, the molecules must be in the same region of the radiation field.

172

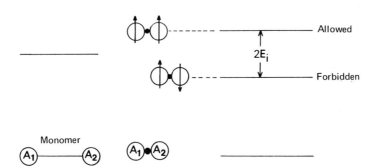

Fig. 6.5. Energy levels of an excited dimer. The arrows point to the direction of the transition dipole. In the illustrated case the transition dipole is perpendicular to the dimer axis. E_i is the dipole interaction energy.

The phase of the electromagnetic wave is, therefore, the same throughout that region, and the generated dipoles must be in phase with each other. If the transition dipoles are aligned with the dimer axis the lower (bonding) state would be the one in which the transition dipoles are in phase and would, hence, be the allowed one. In an oblique arrangement both dipoles have perpendicular and parallel components and both levels are allowed. The energy gap between the levels is twice the dipole interaction energy. The above treatment can be extended to polymers with *n* coupled molecules. In such a case we would have *n* levels which are linear combinations of wavefunctions of the type $\psi_1.\psi_2.\psi_3...\psi_k^*...\psi_n$. Depending on the symmetry of the array, all kinds of spectral phenomena can be predicted: blue shifts, red shifts, narrowed bands, broadened bands, band splitting, etc.

Inductive resonance transfer
In the case of slow transfer, when τ_t is large compared to the period of the lattice vibrations, a "localized treatment" gives an acceptable approximation. In this case the excitation resides for sufficiently long periods in a particular molecule to allow thermal equilibrium among the vibrational levels. The sensitizer molecule will settle into the lower vibrational states before transferring its energy, and the amount of energy transferred correspond to a transition from these lower vibrational states to the ground state. The same amount of energy is then gained by the acceptor molecule (see Fig. 6.6). Since deexcitation of the sensitizer S follows the same pathway as fluorescence, the rate of transfer must be proportional to the overlap of the fluorescence spectrum of S and the absorption spectrum of the acceptor A. In Fig. 6.7a this is shown for the case in which S and A are identical molecules; Fig. 6.7b shows the case in which S and A are dissimilar. One can see that heterogeneous slow transfer can be much more efficient than homogeneous slow transfer.*

T. Förster (1951) has derived a form for the ratio between the rate of slow transfer ($k_t = 1/\tau_t$) and the rate of fluorescence ($k_0 = 1/\tau_0$). This ratio can

---

* The process should not be confused with the trivial process of reabsorption of emitted light. The efficiency of the latter process is very low.

173

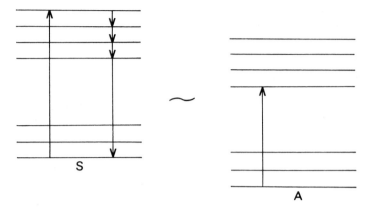

Fig. 6.6. Energy diagram showing the "slow" transfer of excitation energy from a sensitizer molecule S to an acceptor molecule A. Transfer by inductive resonance occurs when the energy lost by S in its deexcitation is gained by A in its excitation.

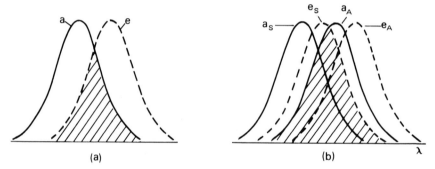

Fig. 6.7. The overlap of the emission spectrum e of S and the absorption spectrum a of A which determines the efficiency of energy transfer by inductive resonance: (*a*) when S and A are similar; (*b*) when S and A are different.

be represented by

$$k_t/k_0 = (R_0/R)^6 \qquad (6.4)$$

in which R_0 is a function of the overlap integral of fluorescence and absorption spectra and the mutual orientation of the transition dipoles, and R is the distance between the interacting molecules.* R_0 has the dimension of length and can be defined as the distance between two molecules at which the rate of slow transfer equals the rate of fluorescence. This value is a measure of the transfer efficiency. L. Duysens (1952) has calculated an R_0 of 69 Å for *in vivo* homogeneous transfer in chlorophyll *a* and 70 Å for the transfer from chlorophyll *b* to chlorophyll *a*. Since the average distance between the molecules in a photosynthetic unit containing chlorophyll *a* and chlorophyll *b* molecules is about 17 Å, this amounts to a very high transfer efficiency. The main physical

---

* $R_0^6 = C \kappa/n^4 \int F(\nu)\epsilon(\nu)/\nu^4 \, d\nu$, in which C is a constant, κ a function of the relative orientation of the dipoles, $F(\nu)$ the spectral emission and $\epsilon(\nu)$ the molar extinction coefficient.

174

reason for this is the unusually good overlap of the absorption and the fluorescence spectra of chlorophyll.

Trapping

In a photosynthetic unit the excitation energy is trapped at a specialized site. Trapping requires that the excitation becomes fixed in one molecule through entry into a localized excited state. Two situations can be visualized in which trapping can occur (see Fig. 6.8). One case is when the trapping molecule T has an excited singlet state which is lower than the excited singlet state of the transferring molecules M. If we consider only slow transfer we can see immediately that the T must act as an energy sink because the overlap of its absorption spectrum and the fluorescence spectrum of the M is greater than the absorption-fluorescence overlap of the M. The trapping efficiency is at least as high as the probability of excitation reaching T (Fig. 6.8a). This situation may prevail in one of the photosynthetic systems of higher plant photosynthesis (see below), where chlorophyll a molecules transfer excitation energy to a specialized chlorophyll, P700, in the reaction center. In this system the transferring chlorophyll a molecules have an absorption

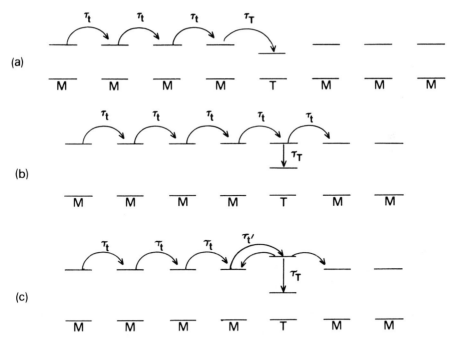

Fig. 6.8. Mechanisms for the trapping of excitation energy from an ensemble of molecules M by a trapping molecule T; (*a*) when the first excited state of T is lower than that of the transferring molecules M; (*b*) when the first excited states of M and T are equal but T has, in addition, an excited state below the first excited state which is absent in M; (*c*) when the first excited state of T is higher than that of M but T has in addition an excited state below the first excited state of M which is absent in M. In cases (*b*) and (*c*) efficient trapping can occur when the trapping time τ_T is far less than the transfer time τ_t.

band that peaks at 680 mm, while the peak of the absorption band of trap molecules is at 700 nm. The energy difference is, thus, about 420 cm^{-1} (= 0.05 eV = 1.15 kcal/mole).

In the second situation the singlet level of T is not different from that of the M, but another excited state (for instance an $n\pi^*$ state, a triplet state, or a charge transfer state) peculiar to T can be entered through the singlet state (Figure 6.8b). Efficient trapping can occur if the trapping time τ_T is much smaller than the transfer time τ_t. This situation may occur in most photosynthetic bacteria in which the peak of the absorption bands of the trapping chlorophyll is the same as that of the transferring chlorophyll. One species of photosynthetic bacteria is known (*Rhodopseudomonas viridis*) for which the wavelength of the peak of the trapping chlorophyll is even lower than that of the transferring chlorophyll. In this case the transfer to the traps is less efficient than the back transfer from the traps to the transferring chlorophyll but once the trapping molecule is excited trapping will be the result when $\tau_T \ll \tau_t$ (Fig. 6.8c).

The total time needed for the migration of excitation energy and trapping $<t>$ is equal to τ_t multiplied by the number of M $\rightarrow$ M transfers required for the excitation to reach T (if $\tau_T \ll \tau_t$). If the trapped state is nonfluorescent the fluorescence yield of the ensemble cannot be greater than $<t>/\tau_0$. Thus, the quenching of the fluorescence in the ensemble is often taken as evidence for the transfer to trapping centers (see below).

The existence of traps can be a result of a variety of effects. An energy level in a trapping molecule could be depressed, thus forming an energy sink by environmental factors. A lowering of a $\pi\pi^*$ level can occur through distortion of the π and the π^* orbitals by local electric fields or electron orbital overlap. An $n\pi^*$ level could be lowered by an electric attraction that encourages the charge displacement involved in the transition. The polarity of the environment (aqueous or lipid phase) can have an influence on the relative position of the lowest $\pi\pi^*$ and $n\pi^*$ levels. It is known that in simple molecules the relative position of these levels can be inverted by transferring the molecule from a nonpolar to a polar solvent.

Chlorophyll fluorescence yield changes

As we already pointed out above, the transfer of excitation energy from the antenna pigments to the reaction center can be followed by observing the changes of the fluorescence yield of the antenna pigments occurring during the transfer. When a reaction center is excited by a transfer of excitation energy from the antenna pigment aggregate (usually followed by a chemical change), the reaction center is inaccessible to following excitation quanta until the original (ground) state of the reaction center is restored. From Eqs. 3.27 and 3.28 (section 3.3) it immediately follows that at high intensities (when many reaction centers are excited at the same time) the fluorescence yield φ_f must be high, and at low light intensities (when many reaction centers are still "open" for excitations) the fluorescence yield φ_f must be low (one of the k_i's can be seen as the rate constant for energy transfer, ultimately

176

to the reaction centers). The variations of the fluorescence yield of an antenna pigment can, thus, give information about the oxidation-reduction processes in the reaction center. This, of course, is only possible when there is an antenna pigment that does fluoresce. Even when the redox reactions in a reaction center cannot be followed directly, by absorption difference spectroscopy, the kinetics of the changes in the fluorescence yield of chlorophyll could reflect the kinetics of those redox reactions.

In photosynthetic bacteria the redox reactions of the reaction center can be followed directly by observing the spectral changes due to the oxidation of the reaction center bacteriochlorophyll and indirectly by looking at the changes of the fluorescence yield of the antenna bacteriochlorophyll a. This fortunate situation has allowed the establishment of a relation between the absorbance changes and the changes of the fluorescence yield, at least for the steady-state situation at different intensity levels of exciting light. From Eqs. 3.27 and 3.28 (section 3.3) it follows that

$$\varphi_f = \frac{k_0}{k_0 + k_i + k_t C_P} \tag{6.5}$$

if it is assumed that the rate constants for the deexcitation processes are of first order and that the trapping rate of the reaction center is proportional to the concentration C_P of "open" reaction centers. In this equation the rate constants k_0, k_i and k_t are those for fluorescence, radiationless transitions, and trapping, respectively. The concentration of the "open" traps C_P is related to the absorption change ΔA according to

$$\Delta A = K_1 + K_2 C_P \tag{6.6}$$

where K_1 and K_2 are constants related to the rate constants and the molar extinction coefficients of the reaction center bacteriochlorophyll in reduced and oxidized form. It is now easy to demonstrate that the inverse of the fluorescence yield is linearly related to the absorption change:

$$1/\varphi_f = \alpha + \beta \Delta A \tag{6.7}$$

in which α and β are constants derived from the constants K_1 and K_2. This linear relationship has been experimentally verified for steady-state conditions at different intensities of light. Figure 6.9 shows the result of such an experiment.

A great deal of the biophysical research in photosynthesis makes use of this technique. It is based, as we have shown, on the fact that changes in the fluorescence yield always reflect changes in the photochemistry as long as the latter starts from the fluorescent state, irrespective of whether or not the chemistry can be measured.

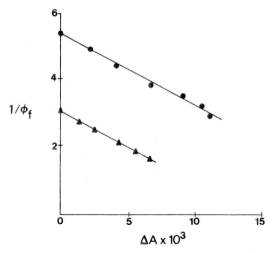

Fig. 6.9. Plots of the inverse fluorescence yield φ_f of bacteriochlorophyll as a function of the extent of the light-induced change of the absorption of the reaction-center bacteriochlorophyll in a suspension of the purple bacterium *Rhodospirillum rubrum* in two different media (from W.J. Vredenberg and L.N.M. Duysens (1963) *Nature* **197**, 355).

6.3 Photosynthetic electron transport in higher plants and algae

The primary reaction

In section 5.1 of Chapter 5 we suggested that photosynthesis essentially occurs in two phases (Eqs 5.1 and 5.2). The first phase is the production, in the presence of light, of a reductant, and the second phase is the utilization, in the absence of light, of this reductant to synthesize sugar by the reduction of CO_2. The light-produced reductant is a pyridine nucleotide, more specifically (for higher plants and algae) NADPH. The light-induced reduction of $NADP^+$ is initiated by so-called *primary reactions* of the type

$$(P.A.) \xrightarrow{h\nu} (P.A)^* \rightarrow (P^+ A^-) \qquad (6.8)$$

in which $h\nu$ is a quantum of light and the asterisk designates an excited state. These reactions occur in the reaction centers, within a time shorter than 10^{-8} s after excitation, through the antenna pigment aggregate. Thus, the primary reaction amounts to the extremely rapid production of a *primary oxidant* (P^+) and a *primary reductant* (A^-).

The two photosystems

In higher plants and algae two such primary reactions are needed to complete the sequence of oxidation-reduction reactions leading to the production of NADPH (*cf.* Eq. 5.1). The sequence, including the two primary reactions, is shown in Fig. 6.10 in which the vertical axis measures the midpoint redox potentials (relative to the standard hydrogen electrode at pH = 7) of

178

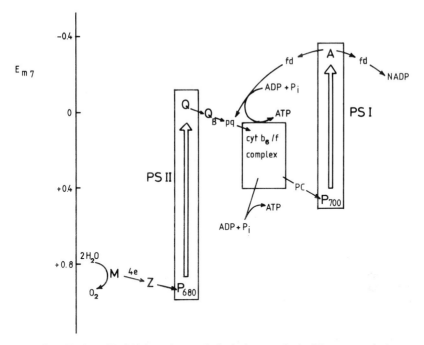

Fig. 6.10. The "Z-scheme" of higher plant and algal photosynthesis. The two vertical arrows represent the electron transport from the primary donor to the primary acceptor in the two reaction center complexes. The primary donors are the "specialized" chlorophyll molecules P680 (Photosystem II) and P700 (Photosystem I). Z is the secondary donor to P680. M is the water splitting enzyme. The primary and the secondary acceptors Q and Q_b of Photosystem II are plastiquinone; pq is plastoquinone of the "pool". The cyt b_6/f complex contains in addition the cytochromes also iron-sulfur centers. PC stands for plastocyanin. The primary acceptor of Photosystem I, A, is presumably an iron-sulfur center protein. fd is ferredoxin (also a nonheme iron protein).

the electron transport mediators involved. The arrows indicate the transfer of electrons. It can be seen from this figure that the primary reaction, labeled Photosystem (PS) II, produces a strong oxidant P680$^+$, capable of oxidizing water and a relatively weak reductant Q$^-$. The other primary reaction, labeled PS I, produces a strong reductant A$^-$ which, through a number of intermediates such as the nonheme iron protein ferredoxin, can reduce NADP$^+$. A relatively weak oxidant P700$^+$ is concomitantly produced in this primary reaction. The weak reductant Q$^-$ of PS II then reacts with the weak oxidant P700$^+$ of PS I through a number of intermediates, such as plastoquinone, cytochromes, and a copper protein, plastocyanin, thus closing the redox chain. The electron transport from Q$^-$ to P700$^+$ is coupled to the formation of ATP in a process called *photophosphorylation*. Based on the difference in redox potential between Q$^-$ and P700$^+$, the coupling is thermodynamically in favor of the production of one ATP molecule for every two electrons transported from Q$^-$ to P700$^+$.

Just like in the respiratory chain, the intermediates between the two primary reactions and the reaction centers themselves are organized in integral membrane protein complexes. The cytochrome b_6/f complex shows many resemblances to complex III of the respiratory chain. The coupling of the electron transport in the complex to photophosphorylation can also be described by Mitchell's chemiosmotic hypothesis (see Chapter 8).

The two reaction centers are each connected to their own antenna system. In fact, the term "Photosystem" applies to the whole of reaction center and antenna system. In higher plants and green algae the PS II antenna system contains chlorophyll a and chlorophyll b. In red algae (like in cyanobacteria, see below) chlorophyll b is replaced by phycobilins. PS I contains only chlorophyll a. Although "spill-over" from PS II to PS I occurs in a limited amount, excitation energy absorbed in one pigment system is usually transferred to its own reaction center. This allows selective excitation of each system; far red light (with wavelengths of 700 nm and larger) is absorbed almost exclusively by PS I while green to orange light (with wavelength between 500 and 650 nm) is absorbed for about 65% by PS II. Far red light, therefore oxidizes all components between the two primary reactions, while green to orange light tends to reduce these components. This has actually been measured by looking at light-induced absorbance changes due to the cytochromes. Results of such experiments have been used as evidence for the existence of the two photosystems.

Cyclic electron transport
The photoreduction of $NADP^+$ is not the only photoreaction which can result from the primary reaction of photosystem I; electrons can also cycle back from the primary reductant A^- to the main chain, thus performing a cyclic electron transport. This cyclic electron transport is coupled to the formation of ATP. In isolated chloroplasts the cycle can be stimulated to a great extent by nonphysiological intermediates such as phenazine methosulfate. The purpose of this light-induced cyclic electron transport may be to provide an additional means for the production of ATP; as we will see below, the production of ATP in the main chain (one ATP molecule per two electrons transported through the chain) is not sufficient for the fixation of carbon in the cyclic series of dark reactions occurring in most higher plants and algae.

Oxygen evolution
In higher plants and algae PS II is directly connected with the splitting (oxidation) of water and the concomitant evolution of oxygen. The development of this oxygen evolving system was crucial for the evolution of life as we know it today. Without it, there would be no atmospheric oxygen and no breathing forms of life.

Much of the early research in photosynthesis was concerned with oxygen evolution. In former days the only way to measure oxygen evolution was by manometric methods. Then, in the 1950's, a new method was developed, which was based on polarography; when oxygen is added to or subtracted from a medium in contact with a negatively charged platinum electrode, an electrical current starts to flow between this electrode and a silver-silver chloride reference electrode. The further development of this method caused much progress in elucidating the kinetics of the process but one most important problem remained: The evolution of one oxygen molecule from water requires the removal of four electrons:

$$2H_2O \rightarrow 4e^- + 4H^+ + O_2 . \tag{6.9}$$

Since only one electron is transported by a "one quantum" primary reaction of the Reaction Center of PS II, the collaboration of four primary reactions is required for reaction 6.9 to proceed. What could be the mechanism by which the transport of one electron at a time in the primary reaction of PS II causes the accumulation of four electric charges?

In the first half of the 1970s this problem was, at least partially, solved by the identification of an enzyme that can accumulate four positive charges. The experiment that led to this identification was carried out by the group of P. Joliot in Paris. Joliot used very sensitive polarographic methods to measure the yield of oxygen from isolated chloroplasts when subjected to a series of very short intense light flashes, after the chloroplasts had been adapted in the dark for a long time. If the flashes are short and intense, one can assume that all reaction centers are excited at the same time. The result of the experiment is illustrated in Fig. 6.11. The yield of the first two flashes turned out to be (near) zero. Only at the third flash a substantial yield was observed. At subsequent flashes, the yield per flash shows an oscillation with a periodicity of *four*; after a few cycles, however, the yield oscillation dies out.

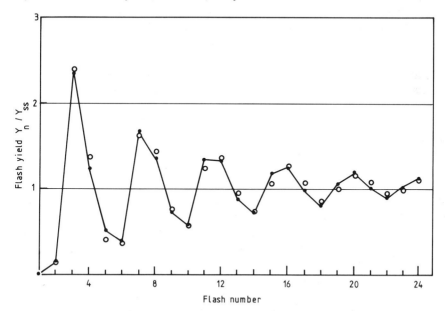

Fig. 6.11. The yield of oxygen per flash in a series of short intense light flashes. The solid points are the experimental data; the open circles are the flash yields calculated from the Kok-Joliot model, assuming 10% "misses" and 5% "double hits". At time zero $S_0:S_1:S_2:S_3 = 0.25:0.75:0:0$ (from B. Forbush, B. Kok, and M. McGloin (1971) *Photochem. Photobiol.* **14**, 307–321).

The explanation of this remarkable result was given by B. Kok, who collaborated with Joliot on this subject: The primary reaction extracts (through a still hypothetical secondary donor to P680, called Z) an electron from an enzyme (or enzyme complex) M. By losing an electron, M changes from one "state" to another, each state characterized by the accumulation of an additional positive charge. The enzyme M thus can be in five "states" which we designate by S. The state with no net charges is S_0, with one charge is S_1 and so on until S_4. In state S_4 the enzyme transfers to S_0 by the release of one oxygen molecule. This "ratchet" mechanism is illustrated in Fig. 6.12.

To explain the fact that, after dark adaptation, the first maximum yield occurs at the *third* and not at the fourth flash, as would be expected from the proposed mechanism, it is assumed that both, the states S_0 and S_1 are stable in the dark and that the states S_2 and S_3 decay, in the dark, to S_1 and not to S_0. Hence, after a long dark period about 1/4 of the enzyme M is in state S_0 and 3/4 of it in state S_1. Furthermore, to explain the damping of the oscillation, it is assumed that some of the excitations do not result at all in a state transformation ("misses") and some others in two transformations at once ("double hits"). This would cause the M-enzyme entities to gradually run out of step. By making assumptions on the relative occurrences of "misses" and "double hits", B. Kok and collaborators were able to predict flash yields that came reasonably close to the experimental results (see Fig. 6.11).

It must be emphasized that, as yet, the proposed model is purely formalistic. As has been mentioned, the enzyme is not yet characterized. The fact that manganese is a requirement for the water splitting function of the PS II reaction center is suggestive, of course. Manganese can accumulate four positive charges. The results of some Mn-ion NMR experiments carried out recently by the

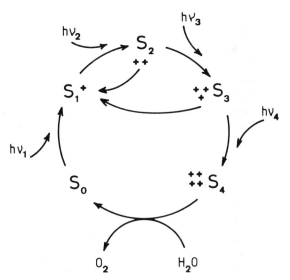

Fig. 6.12. The Kok-Joliot model of oxygen evolution in photosynthesis.

group of Govindjee of the University of Illinois also were suggestive of a role of manganese in the oxygen evolving system. No further details are, as yet, available and much research is still left to be done on this system. On a formal basis, however, the model so far is consistent with the kinetic information on oxygen evolution, which is available to date.

6.4 Photosynthetic electron transport in prokaryotes

Bacterial photosynthesis
At present there is no evidence for more than one type of primary reaction in photosynthetic bacteria. The electron transport reactions set in motion by the primary reaction in these organisms are more primitive, but also more flexible than those in higher plants. Most of the main electron transport chain appears to be cyclic (see Fig. 6.13). The primary reductant Q_1^- generated by the primary reaction, reduces a secondary acceptor Q_2 that, in turn, reduces a quinone pool (ubiquinone in most bacteria). The quinone pool reacts with a cytochrome b/c_1 complex that, again, resembles in many aspects the Complex III of mitochondria. It contains, in addition to cytochrome c_1, a number of b-type cytochromes and an iron-sulfur protein. The more soluble cytochrome c_2 then becomes reduced by the complex and re-reduces the primary donor, $P870^+$, thus closing the cycle. More details of this cyclic electron transport will be discussed in Chapter 8 in connection with the chemiosmotic mechanism of photophosphorylation.

During this cyclic electron transport ATP, or some other "high energy intermediate", is produced and can then be used to drive reactions against the thermodynamic gradient in coupled reactions. It has been demonstrated, for example, that preparations of photosynthetic bacteria can catalyze the reduction of NAD^+ by succinate, a reaction with a positive free energy change. The energy for such a reaction, occurring in darkness, can be obtained from the coupled hydrolysis of ATP or even of pyrophosphate. Such reactions most probably do occur under *in vivo* physiological conditions, so that the main task of the light reaction in photosynthetic bacteria (Fig. 6.13) would be to provide energy intermediates in suitable form (ATP or others). The occurrence, under physiological conditions, of other direct light-driven electron transport reactions still cannot be ruled out, however.

Cyanobacteria
Cyanobacteria are also prokaryotic organisms that carry out photosynthesis. However, they do not resemble photosynthetic bacteria; like higher plants and algae, they can split water, producing oxygen, and reduce $NADP^+$, using two primary reactions in series. Their main pigment is chlorophyll a, just as in higher plants, but they have phycobilins, in particular phycocyanin, as accessory pigment. This explains their blue-greenish color. From a physiological point of view they could be classified among the higher plants and

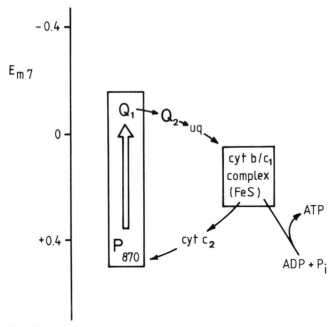

Fig. 6.13. A diagram of the light-induced electron transport in photosynthetic bacteria. The vertical arrow represents the reaction center-associated electron transport from the primary donor P (a "specialized" bacteriochlorophyll *a*) to a primary acceptor (a quinone, usually ubiquinone, sometimes menaquinone). This transport drives a cyclic reaction in which participate a quinone pool (uq), the cytochrome *b/c₁* complex that contains also iron-sulphur centers, and a soluble cytochrome *c₂*. The function of the cyclic electron transport is presumably the production of ATP (or some other form of potential energy) that can be used in coupled reactions to drive the synthesis of cell constituents from substrates.

algae and, therefore, were often called blue-green algae. This situation persisted until the mid 1970s when it was discovered that under certain circumstances a number of species of cyanobacteria (especially those of the genus *Oscillatoria*) are able to short-circuit Photosystem II, using only Photosystem I to grow. Evidently, no water is split and no oxygen is evolved under those circumstances. The substrate they use to grow on is sulfide. In fact, it is sulfide that triggers the transformation from the oxygen evolving (oxygenic) to the sulfide oxidizing (anoxygenic) mode of photosynthesis. The transformation is genetic; in the presence of an antibiotic that inhibits transcription of DNA they fail to make the transformation.

It is tempting to compare the anoxygenic form of photosynthesis in cyanobacteria with that of photosynthetic bacteria. This comparison is not obvious, however. Structurally, the primary reaction in the reaction center of photosynthetic bacteria relates more to PS II in plants than to PS I (see below). In the anoxygenic photosynthesis of cyanobacteria it is PS II which is switched off by the transformation. Nevertheless, the discovery of this form of photosynthesis is important and further research may provide clues to understanding how photosynthesis, and in particular oxygenic photosynthesis, may have evolved. Cyanobacteria are known to have occurred in a very early stage of biological evolution.

6.5 Reaction centers

Definition

The existence of reaction centers was already postulated in the 1930s when the Photosynthetic Unit was defined. The first spectroscopic evidence for it surfaced in the 1950s, when, first, L. Duysens reported light-induced absorbance changes in the near infrared spectrum of a purple photosynthetic bacterium, and, somewhat later B. Kok reported light-induced absorbance changes in the spectral region around 700 nm in chloroplasts. Both types of absorbance changes were interpreted as due to the oxidation of bacteriochlorophyll *a*, respectively chlorophyll *a*; they were reproduced almost exactly by chemical oxidation. It was recognized that in both cases the reacting chlorophyll was not part of the bulk of the pigment molecules, but appeared sequestered in specialized structures. According to the wavelength at which the maximum absorbance change occurred, the pigments were called P870 and P700 respectively. This nomenclature was subsequently extended; due to the extensive variety of spectral forms of bacteriochlorophyll in different species of photosynthetic bacteria, names vary, presently, from P840 (green bacteria), P870 and P880 (*Rhodospirillum*), P890 (*Chromatium*) and P960 (*Rhodopseudomonas viridis*). In the latter species the reaction center chlorophyll is bacteriochlorophyll *b*.

P700 turned out to be chlorophyll *a* in the reaction center of PS I. Absorbance changes due to the light-induced oxidation of the PS II reaction center pigment were much harder to detect. They were finally discovered, some 15 years after the discovery of P700. They were also due to a specialized and sequestered form of chlorophyll *a* that, again in accordance with its absorption maximum, is called P680.

Structure

None of the two reaction centers of higher plant or algal photosynthesis have as yet been isolated in pure form, although preparations containing a limited amount of antenna chlorophylls have been obtained. In contrast, photoactive reaction centers of several species of photosynthetic bacteria have been isolated in pure form. This allowed extensive studies of composition, structure and function. All reaction centers investigated so far are protein complexes consisting of three subunits, the L (for light), the M (for medium), and the H (for heavy) subunits, having molecular weights of about 21,000, 24,000 and 28,000 respectively, sometimes bound to a fourth cytochrome subunit. Two of these, the L and M units contain the pigments and are together the photoactive part of the complex. Figure 6.14a shows the spectra of a reaction center preparation, containing the L and M subunits, in the reduced (drawn

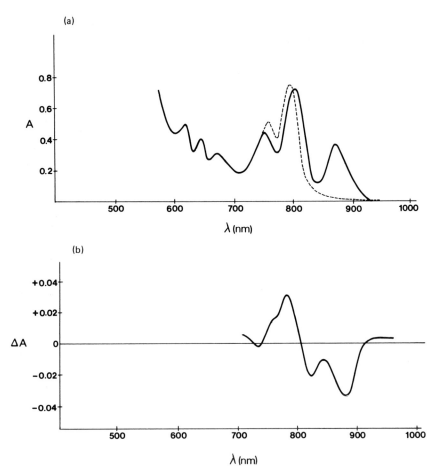

Fig. 6.14. (*a*) Spectra of the reduced (drawn line) and the oxidized (dashed line) form of a reaction center preparation from the photosynthetic bacterium *Rhodospirillum rubrum*. (*b*) The light-minus-dark difference spectrum of the same preparation.

line) and the oxidized (dashed line) state. Figure 6.14b is the spectrum of the difference between an illuminated sample and a sample kept in the dark. This difference spectrum corresponds exactly to the difference of the two spectra shown in Fig. 6.14a.

There are always four bacteriochlorophyll molecules per reaction center, two of which forming a "special pair". In addition, there are two bacterio-pheophytin molecules (pheophytin is a chlorophyll without the central magnesium atom), one (or sometimes two) quinone (ubiquinone or mena-quinone, depending upon species), and one iron atom. The structure of the reaction center from *Rhodopseudomonas viridis* has recently been elucidated by X-ray diffraction studies which largely confirm earlier predictions about the relative orientations of the chromophores. The structure is discussed in section 3.7 of Chapter 3. The two bacteriochlorophyll molecules forming the "special pair" appear to interact particularly strongly with each other, almost constituting a dimer. It is this special pair that becomes photooxidized in a very short time after excitation.

The primary reactions in the RC complex

Figure 6.15 shows the pathway followed by the electron on an approximate energy scale. The lowest excited singlet state of the "special pair", P, is about 1.38 eV above the ground state. In the oxidized dimer the spin of the unpaired electron is delocalized over the π-systems of the two chlorophyll molecules. The excitation of the dimer causes the transfer of an electron to an intermediate acceptor I, a step which takes at most a few picoseconds. The electron then is transferred to the primary acceptor Q_a and subsequently to the secondary acceptor Q_b, both quinones. The back reactions take the times given in the figure when further electron transport is blocked either by chemical reduction of the successor or by specific inhibitors.

The transient reduction of the early electron acceptor I was detected from the difference spectra of optical absorbance changes following a very short flash (see Figure 6.16). The difference spectrum includes bleaching at 545, 760, and 800 nm, a broad absorbance increase near 670 nm and a small bleaching at 600 nm that can be seen when the reaction centers are excited at 530 nm. The absorption bands at 545 and 760 nm can be attributed to bacteriopheophytin; I appears to be one of the pheophytins present in the reaction center. The absorption bands at 600 and 800 nm are due to bacteriochlorophyll. These changes suggest that I is a complex of pheophytin and bacteriochlorophyll but this still is a controversial matter. The quantum yield of I^- appears to be similar to that of P^+, that is near unity, which suggests

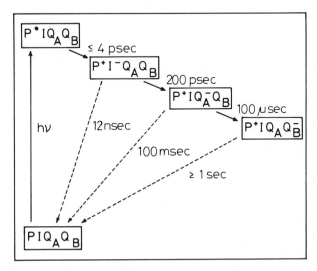

Fig. 6.15. A schematic representation of the early steps of the electron transport in the reaction center of photosynthetic bacteria (from W.W. Parson and B. Ke (1982), in Govindjee (Ed.) "Photosynthesis" Vol. I, Acad. Press, New York, p. 331–385.

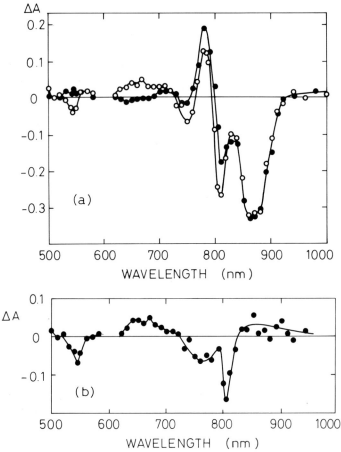

Fig. 6.16. Difference spectra of absorbance changes resulting from excitation of a reaction center preparation from the photosynthetic bacterium *Rhodobacter sphaeroides* with 8 psec flashes at 600 nm. a. Absorbance changes measured at 20 psec (open circles) and at 3 nsec (closed circles) after excitation; b. difference between the absorbance changes at 20 psec and at 3 nsec, showing peaks characteristic for bacteriopheophytin (from C.C. Schenck *et al.* (1981), *Biochim. Biophys. Acta* **635**, 383–392).

that the electron transport is a step in the normal photosynthetic electron transfer.

If I involves both bacteriopheophytin and bacteriochlorophyll the question must be answered whether only one of them undergoes a reduction or whether the electron is shared by both molecules. Evidence obtained with electron spin resonance techniques could be interpreted most simply by the view that the unpaired electron is not shared but is localized on the pheophytin. It seems likely that bacteriochlorophyll would interact more directly with the special pair and serves as the initial electron donor to pheophytin. Moreover, thermodynamically, electrons from bacteriochlorophyll to pheophytin go "downhill".

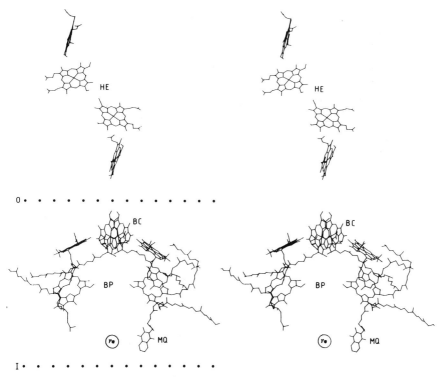

Fig. 6.17. A stereo drawing of the chromphores of the reaction center of the photosynthetic bacterium *Rhodopseudomonas viridis* lifted out of the reaction center structure (cf. Fig. 3.52, Chapter 3) retaining their relative position in the structure; HE are the heme groups of the cytochrome subunit; BC is bacteriochlorophyll, BP is bacteriopheophytin, and MQ is menaquinone (courtesy of Drs. H. Michel and J. Deisenhofer).

The structure of the reaction center of the photosynthetic bacterium *Rhodopseudomonas viridis* is shown in Fig. 3.52 (Chapter 3). Figure 6.17 is a stereo drawing of the chromophores shown in the positions they have in the structure. It appears that photo-induced charge separation occurs only in that part of the reaction center (the left side in Fig. 6.17) in which the pheophytin is adjacent to the primary quinone. Whether the other side, containing a bacteriochlorophyll and a bacteriopheophytin, has any function at all is unknown and seems unlikely at the time of this writing.

Reaction centers of higher plants and algae
There is evidence that the reaction center of PS II in higher plants and algae has a similar structure. P680 seems to be also a "special pair" of chlorophyll *a* and the reaction center contains pheophytin and (plasto-)quinone. A direct structure determination still awaits isolation in pure form without antenna pigments and crystallization.

As yet, little is known about the structure and the primary reactions of the reaction center of PS I. At present, it seems unlikely that the chlorophyll *a* in the PS I reaction center also is in the form of a special pair. Most probably, the primary acceptor is a ferredoxin, a protein containing nonheme iron.

The secondary quinones

In both higher plants and bacteria the secondary quinone (Q_b) may serve as a "gate" to deliver electrons for the reduction of the quinone pool. Quinone in the reduced form is stable only when it is fully hydrated, thus requiring two electrons and two protons. Each primary reaction delivers only one electron at a time. The "gate" would act as some sort of a buffer for two electrons which could function as follows. A first primary reaction transports an electron via Q_a to Q_b:

$$(Q_aQ_b) \overset{hv}{\to} (Q_a^-Q_b) \to (Q_aQ_b^-) \qquad (6.10a)$$

after which a second primary reaction transports the second electron:

$$(Q_aQ_b^-) \overset{hv}{\to} (Q_a^-Q_b^-) \to (Q_aQ_b^{2-}). \qquad (6.10b)$$

In this state the quinone pool q can be reduced in the dark:

$$(Q_aQ_b^{2-}) + q + 2\,H^+ \to (Q_aQ_b) + qH_2. \qquad (6.10c)$$

If such a mechanism were operative, the flash-induced reduction of the quinone pool q should show a periodicity of two. The redox state of the quinone pool can be measured indirectly by a number of methods and a periodicity of two indeed was measured by subjecting dark-adapted chloroplasts to a series of flashes.

6.6 Carbon fixation

The Calvin cycle

The dark stage of photosynthesis is the fixation of CO_2, a process which occurs in the dark. The formation of sugar, and eventually of other cell constituents, results from a cycle of chemical reactions in which CO_2 is incorporated into a five-carbon compound and the resulting unstable six-carbon compound is split into two three-carbon compounds which are subsequently reduced. The *Calvin cycle*, as it is often called after its discoverer Melvin Calvin, is given in a schematic and somewhat simplified form in Fig. 6.18. The five-carbon compound ribulose bisphosphate picks up a molecule of carbon dioxide in a reaction that is catalyzed by the enzyme *ribulose bisphosphate carboxylase* and the resulting six-carbon compound immediately breaks down into two molecules of the three-carbon compound 3-phosphoglyceric acid. The phosphoglyceric acid is then phosphorylated (by ATP) and reduced (by NADPH) into glyceraldehyde 3-phosphate (which is a triose phosphate). A sixth of the glyceraldehyde is then converted into glucose through a series of enzymatic reactions. The remaining part goes through a cycle of reactions which includes one more ATP-mediated phosphorylation step, to form again

190

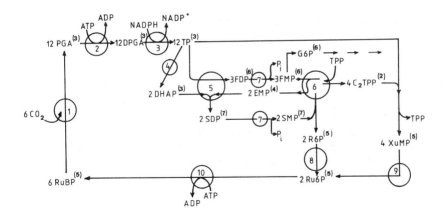

Substrates

		Enzymes	
RuBP	Ribulose bisphosphate	1	Ribulose bisphosphate carboxylase/oxygenase
PGA	phosphoglyceric acid	2	phosphoglyceric acid kinase
DPGA	1,3-diphosphoglyceric acid	3	triose phosphate dehydroxykinase
TP	glyceraldehyde 3-phosphate	4	triose phosphate isomerase
DHAP	dehydroxyacetone phosphate	5	aldolase
FDP	fructose 1,6-diphosphate	6	transketolase
FMP	fructose 6-phosphate	7	diphosphatase
G6P	glucose 6-phosphate	8	phosphoribose isomerase
EMP	erythrose 6'6-phosphate	9	ribulose phoshate-xylulose isomerase
SDP	seduheptulose 6-1,7-diphosphate		
SMP	seduheptulose 7-phosphate	10	ribulose 5-phosphate kinase
R6P	ribose 5-phosphate		
Ru6P	ribulose 5-phosphate		
XuMP	xylulose 5-phosphate		
TPP	thyamine (vitamine B1)		

Fig. 6.18. The Calvin cycle.

the five-carbon compound ribulose bisphosphate. Thus, to "grind out" one molecule of glucose the cyclic reaction "mill" has to carboxylate and dismutate six molecules of ribulose bisphosphate through the formation of twelve molecules of phosphoglyceric acid. These twelve molecules need twelve ATP molecules and twelve NADPH molecules to form twelve glyceraldehyde phosphate molecules. Two of the glyceraldehyde phosphate molecules then join to form ultimately one glucose molecule. The other ten glyceraldehyde phosphate molecules go through the cycle to form six new molecules of ribulose bisphosphate, utilizing six more ATP molecules. The total process, thus, needs twelve molecules of NADPH and eighteen molecules of ATP; in other words, for each two molecules of $NADP^+$ reduced in the light reaction, three ATP molecules are needed to perform the Calvin cycle as described. An extra

191

photophosphorylation site in a light-induced cyclic electron pathway would fulfill this need.*

Photorespiration

The enzyme ribulose bisphosphate carboxylase is not entirely specific for CO_2 but can also accept oxygen as a substrate. In that case ribulose bisphosphate is converted into one molecule of 3-phosphoglyceric acid and one molecule of phosphoglycollic acid. The latter is further metabolized outside the chloroplast with the release of CO_2. Thus, this pathway consumes oxygen in a light-dependent pathway and liberates CO_2. Therefore it is called *photorespiration*, although it has nothing to do with the respiratory oxygen uptake; for one thing, it dissipates energy instead of conserving it. The significance of this process is not quite understood. It reduces the quantum yield of carbon fixation which can be a handicap when CO_2 levels are low.

Plants have learned to cope with this apparent shortcoming of the enzyme by producing a lot of it. Ribulose bisphosphate carboxylase is the most abundant protein in plants. Moreover, there are a number of ancillary pathways to keep the CO_2 level in the stroma of the chloroplast high.

The C4 pathway

One such a pathway is characteristic of a number of plants, such as tropical grasses, maize and sugar cane. In these plants CO_2 is first assimilated, not into phosphoglyceric acid but into four-carbon dicarboxylic acids, hence the name *C4-plants*. The C4 pathway takes place in a special kind of cells, the *mesophyll cells*. In these cells pyruvate is converted into phosphoenolpyruvate at the expense of two molecules of ATP. Then, CO_2 is assimilated, converting the phosphoenolpyruvate into oxaloacetate that, in turn, is converted into malate. The malate migrates to another group of cells which are adjacent to the tubular structures that carry the products of photosynthesis out of the leaf. In these so-called *bundle sheath cells* the malate is decarboxylated into pyruvate, releasing CO_2 which then can be used in the Calvin cycle. In such a way, the plant can have the disposal of a reservoir of "bound" CO_2. This could become necessary when in hot sunny weather the stomata are closed, preventing CO_2 to enter the leaf. These stomata, which are little orifices on the surface of leaves, close in daytime to reduce water loss. But in daytime there is plenty of light to produce ATP that is required for the C4 pathway.

---

* Several enzymes of the Calvin cycle are activated by light. This light-activation occurs when photoreduced ferredoxin reduces sulfur bonds (S-S) to 2SH, thus changing the conformation of the enzyme into an active form. In the dark, the SH-groups are reoxidized by molecular oxygen. Since the activation is light-dependent and reverted in the dark, it provides for a means of control of the cycle (see B.B. Buchanan, 1980).

Bibliography

Buchanan, B.B. (1980). Role of Light in the Regulation of Chloroplast Enzymes, *Ann Rev. Plant Physiol.* **31**, 341–374.

Cséke, C., and Buchanan, B.B. (1986). Regulation of the Formation and Utilization of Photosynthate in Leaves, *Biochim. Biophys. Acta, Reviews on Bioenergetics* **853**, 43–63.

Duysens, L.N.M. (1952). "Transfer of Excitation Energy in Photosynthesis". Thesis, Univ. of Utrecht, DUM v/h Kemink en Zn. N.V., Utrecht.

Förster, TH. (1951). "Fluoreszenz Organische Verbindungen", VandenHoeck and Ruprecht, Göttingen.

Govindjee, (Ed.). (1982). "Photosynthesis", Vol. I and II, Acad. Press, New York.

Joliot, P. and Kok, B. (1975). Oxygen Evolution in Photosynthesis, in "Bioenergetics of Photosynthesis" (Govindjee, Ed.), pp. 387–412, Acad. Press, New York.

Kasha, M. (1963). Energy Transfer Mechanisms and the Molecular Exciton Model for Molecular Aggregates, *Rad. Res.* **20**, 55–71.

Kok, B., and Hoch, G. (1963), in "Light and Life" (W.D. McElroy and B. Glass, Eds.), pp. 397–416, Johns Hopkins Press, Baltimore.

7. Biological transport processes

7.1 Passive and active transport

Transport across membranes

Transport processes are an integral part of biological function. For example, the energy converting processes which we have discussed in the previous chapters need a continuous supply of substrates and a continuous disposal of products and waste. It is evident that there can be no respiration when there are no means for oxygen and substrates (glucose) to penetrate the cells and the organelles; carbon dioxide has to be removed as well. Often, ATP produced at one point in the cell must be transported to another. Many other substances, neutral as well as charged, have to be transported in order to make vital processes function.

Compartmentalization seems to be the structural feature by which cells carry out their function. This is more conspicuous in the higher developed and differentiated eukaryotic cells than it is for the prokaryotes, although it seems essential for these more primitive cells as well. Compartmentalization is accomplished by membranes, and where some transport of matter occurs through channels bordered by membranes (the endoplasmic reticulum and the Golgi apparatus, for example) the *selective* transport often occurs through the membranes themselves. By passive and active transport the chemical integrity inside the compartments of the cell and the cell organelles is kept constant within narrow limits, thus providing optimal conditions for the life processes. By *passive* transport we mean diffusion in the direction of the thermodynamic gradient; *active* transport is the movement of solutes against the thermodynamic gradient. The latter requires an energy source and mechanisms to couple the energy input to the transport. The selectivity is a consequence of the permeability of the membrane itself, often determined by the particular molecular mechanism of the transport. Membrane transport and permeability are the subjects of this chapter.

Active transport

In cells or cell organelles large differences in the concentrations of (charged or uncharged) solutes between the inside and outside of the membrane-surrounded vesicle can be found, even when the membrane is permeable to such solutes. In many cells, for example, the cytoplasmic membrane is perfectly permeable to K^+ and Na^+. The concentration of K^+ inside the cell, however, is many times higher than the concentration of the ion in surrounding medium, while Na^+ has a lower concentration inside than it has outside. One would say that a situation like this can only be maintained by active transport of K^+ to the inside and of Na^+ to the outside. This is, indeed, true for the cytoplasmic membrane of most vertebrate cells; it contains an enzyme, the so called $K^+Na^+ATPase$, which takes care of the transport of the two ions in *antiport* (each in opposite directions) at the expense of the hydrolysis of ATP.

The transport of Na^+ and K^+ in antiport is an example of *coupled transport*. Antiport is coupled transport in opposite directions. Coupled transport in the same direction is called *symport*. There are many antiport and symport transporters of charged and uncharged substances in different kinds of membranes. The mitochondrial membrane, for instance, contains enzymes that couples the transport of phosphate to that of dicarboxylate, malate to that of citrate, Na^+ to that of H^+, ATP^{4-} to that of ADP^{3-} and many others. Some of these coupled transports are not electroneutral, such as the ATP^{4-}/ADP^{3-} and the Na^+/K^+ antiports.

7.2 Osmotic equilibrium

Osmotic pressure

There are situations in which a concentration gradient of a solute, charged or uncharged, can be maintained at equilibrium. This occurs when the membrane has semipermeability characteristics. A situation in which this occurs is the following: Consider a membrane separating two compartments I and II (Fig. 7.1). Compartment I contains a solvent S, while in compartment II a solute A is dissolved in the solvent S. Let the membrane only be permeable to the solvent. The molecules of the solvent tend to move from compartment I to compartment II because of the concentration difference. Net movement in equilibrium, however, is counteracted by a buildup of pressure in compartment II. This can be translated into thermodynamical terms as follows.

The change of free energy when a mole of solvent moves from compartment I to compartment II is

$$\Delta G = \mu_S^{II} = \mu_S^{I} \tag{7.1}$$

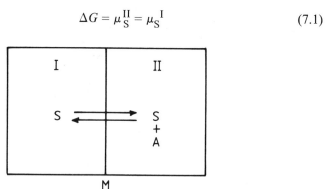

Fig. 7.1. Two compartments I and II, separated by a semipermeable membrane permeable to the solvent S but no to the solute, A.

in which μ_S^I and μ_S^{II} are the chemical potentials of the solvent in compartments I and II, respectively. At equilibrium $\Delta G = 0$. Using the expressions for the chemical potential given in terms of the mole fraction x_S (the ratio between the number of moles of the solvent and the total number of moles of solvent and solute) Eq. 7.1 becomes

$$\mu_S^{\circ II} + RT \ln x_S^{II} - \mu_S^{\circ I} - RT \ln x_S^I = 0. \tag{7.2}$$

This reduces to

$$\mu_S^{\circ II} - \mu_S^{\circ I} + RT \ln x_S^{II} = 0. \tag{7.3}$$

since $x_S^I = 1$ because there is no solute in compartment I. The μ_S° depend only on pressure. In order to find this dependence we can apply the Gibbs equation and the defining differential of the Gibbs free energy to the solvent (see Appendix II). This yields

$$d\mu_S = v_S dP \tag{7.4}$$

in which v_S is the molar volume of the solvent. Integrating Eq. 7.4 between the appropriate limits,

$$\int_{\mu_S^{\circ I}}^{\mu_S^{\circ II}} d\mu_S = \int_{P^I}^{P^{II}} v_S \, dP$$

yields

$$\mu_S^{\circ II} - \mu_S^{\circ I} = v_S(P^{II} - P^I) \tag{7.5}$$

if the solvent is assumed to be incompressible, which in our case is a reasonable assumption. Substituting Eq. 7.3 into Eq. 7.5, and rearranging the terms, gives

$$P^{II} - P^I = \pi = -\frac{RT}{v_S} \ln x_S^{II} \tag{7.6}$$

defining $P^{II} - P^I = \pi$ as the *osmotic pressure*. We can express the osmotic pressure in terms of the solute concentration by putting $x_S = 1 - X_A$ in which X_A is the total mole fraction of solute. For dilute solutions X_A is small with respect to x_S and we can make the approximation

$$X_A = \frac{n_A}{n_S + n_A} \approx \frac{n_A}{n_S}. \tag{7.7}$$

Since $n_S v_S$ is the solvent volume V we have

$$\frac{n_A}{n_S} = v_S C_A \tag{7.8}$$

in which C_A is the concentration of the solute. Furthermore, the logarithm of Eq. 7.6 can be expanded to yield

$$\ln(1 - X_A) = - X_A - \frac{X_A{}^2}{2} - \frac{X_A{}^3}{3} - \ldots \qquad (7.9)$$

in which, again for dilute solutions, the higher-order terms can be neglected. Using these approximations, Eq. 7.6 can be reduced to

$$\pi = RTC_A \qquad (7.10)$$

which is the well-known Van 't Hoff equation.

7.3 Ionic equilibrium

The Nernst potential
The osmotic pressure can thus be seen as raising the chemical potential of the solvent in the solution to that of pure solvent as required. A similar situation exists when we have a charged solute. Suppose that the membrane separates two compartments I and II with different concentrations of an electrolyte C^+A^- and that it is permeable only to ions of one sign, for instance, the cations C^+. At equilibrium, the change in the free energy when a mole of these cations pass from compartment I to compartment II is zero:

$$\Delta G = \tilde{\mu}_C^{II} - \tilde{\mu}_C{}^I = 0 \qquad (7.11)$$

in which the $\tilde{\mu}_C$ are the *electrochemical* potentials consisting of a chemical component μ and an electrical component $z\mathscr{F}\psi$ (with z as the valence of the ionic species, $\mathscr{F}$ as the Faraday constant, and ψ as the electrical potential). Substituting the appropriate expressions for the electrochemical potentials in 7.11 we obtain

$$\mu_C{}^\circ + RT \ln c_C{}^I + z_C \mathscr{F} \psi^I = \mu_C{}^\circ + RT \ln c_C^{II} + z_C \mathscr{F} \psi^{II}. \qquad (7.12)$$

Solving for the electrical potential difference gives

$$\psi^{II} - \psi^I = \frac{RT}{z_C \mathscr{F}} \ln \frac{c_C{}^I}{c_C^{II}} \qquad (7.13)$$

The electrical potential difference is, thus, proportional to the logarithm of the ratio of the two concentrations. For cations z is positive and the electrical potential is higher on the more dilute side of the membrane. Equilibrium is attained because the buildup of electrical potential on the dilute side of the membrane raises the electrochemical potential of the solution in the more diluted compartment to that of the more concentrated solution in the other compartment. If the membranes were only permeable to anions, we would have had the reverse situation. It is to be noted that in each compartment (I and II) the law of electrical neutrality is still valid, because the charge difference (or charge displacement) cannot be detected; it is only manifest as an electrical potential difference. Such a potential difference is often referred to as a *diffusion potential* (because it results from an apparent diffusion of ions of one sign through the membrane). Eq. 7.13 often is called the *Nernst equation*. If an external electrical field is applied to a membrane which is permeable to ions of one sign only, and the membrane is separating two compartments each containing solutions of the ion, the concentrations of the ion at equilibrium are given by the Nernst equation.

Donnan equilibrium
A particular case of ionic equilibrium across membranes, such as we have just described, is the Donnan equilibrium. In this case an electrical potential exists at equilibrium even when the membrane is permeable to (relatively small) ions of both signs. This happens when one of the two compartments separated by the membrane contains, in addition to a salt to which the membrane is permeable, a large molecule (a protein, for example) bearing a net charge, to which the membrane is not permeable (Fig. 7.2). Suppose that compartment I contains a solution of a simple univalent electrolyte C^+A^- and that compartment II contains a solution of the same electrolyte together with a protein salt P, at a concentration c_P bearing a net charge z_P. The membrane separating the two compartments is permeable to both ions of the electrolyte. Suppose, furthermore, that an appropriate osmotic pressure exists such that the chemical potential of the solvent is equal in both compartments. At equilibrium the free energy change when a mole of the simple electrolyte is transferred from one compartment to the other is zero. This means that

$$\mu^I_{C^+A^-} = \mu^{II}_{C^+A^-} \tag{7.14}$$

or, using the appropriate expressions for the potentials,

$$\mu^{\circ I}_{C^+A^-} + RT \ln c^I_{C^+} c^I_{A^-} = \mu^{\circ II}_{C^+A^-} + RT \ln c^{II}_{C^+} c^{II}_{A^-} \tag{7.15}$$

If we ignore the small effects of the pressure difference on the standard potentials of the salt in the two compartments then

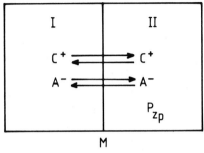

Fig. 7.2. Two compartments I and II separated by a membrane permeable to the small ions C^+ and A^- but not to a large charged molecule P_{z_P}.

$$\mu^{\circ I}_{C^+A^-} = \mu^{\circ II}_{C^+A^-} \tag{7.16}$$

and it follows that

$$c^I_{C^+}c^I_{A^-} = c^{II}_{C^+}c^{II}_{A^-}$$

or

$$c^I_{C^+}/c^{II}_{C^+} = c^I_{A^-}/c^{II}_{A^-} = r \tag{7.17}$$

The ratio r is called the Donnan ratio.

The law of electrical neutrality dictates that, in compartment I,

$$c^I_{C^+} = c^I_{A^-} \tag{7.18}$$

and in compartment II,

$$c^{II}_{C^+} = c^{II}_{A^-} - z_P c_P \tag{7.19}$$

From (7.17), (7.18), and (7.19) it follows that

$$r^2 = c^{II}_{A^-}/c^{II}_{C^+} = c^{II}_{A^-}/(c^{II}_{A^-} - z_P c_P) \tag{7.20}$$

Equation 7.20 shows us that if the net charge on the protein is negative ($z_P < 0$), $r^2 < 0$ and, hence, $r < 1$. Consequently, according to 7.17, $c^{II}_{A^-} < c^I_{A^-}$ and $c^I_{C^+} < c^{II}_{C^+}$. The Nernst equation, Eq. 7.13, tells us then that there must be a negative electrical potential,

$$\psi^{II} - \psi^I = \frac{RT}{\mathscr{F}} \ln \frac{c^{II}_{A^-}}{c^I_{A^-}} = \frac{RT}{\mathscr{F}} \ln \frac{c^I_{C^+}}{c^{II}_{C^+}} \tag{7.21}$$

across the membrane. According to 7.17 this potential is proportional to the logarithm of the Donnan ratio:

$$\psi^{II} - \psi^I = \frac{RT}{\mathscr{F}} \ln r. \tag{7.22}$$

When the net charge on the protein is positive ($z_P > 0$), the electrical potential evidently is positive.

The above derivation was made for a simple univalent salt. It is relatively easy to show, however, that for polyvalent electrolytes one can define a Donnan ratio for each salt k

$$r = (c_k^I / c_k^{II})^{1/z_k} \tag{7.23}$$

in which the subscript k refers to the kth ion species with a charge z_k. Equation 7.22 is, thus, applicable for the general case.

200

7.4 Flow across membranes

Flow equations
So far only equilibrium situations in which no net flow of matter occurs have been discussed. The equilibrium, of course, can be a dynamic one, describing a steady state in which the flow in one direction equals the flow in the opposite direction. Situations in which there is a net flow of matter can be described by transport equations of the kind

$$J = -L \text{ grad } \mu \tag{7.24}$$

or, in the one-dimensional case,

$$J = -L \, d\mu/dx \tag{7.25}$$

in which J is the flow, or the amount of matter passing a unit area in a unit time, L is a coefficient related to the mobility of matter through the medium (a phenomenological coefficient), and μ is a potential function. We can apply this to simple diffusion of a solute in solution in one dimension. If v_i is the average velocity of a mole of solute i, the flow, in moles per unit area per unit time, is

$$J_i = c_i v_i \tag{7.26}$$

in which c_i is the concentration of solute i in moles per unit volume. The velocity v_i is proportional to a force F, which is equal and opposite to the driving force:

$$v_i = \omega_i F \tag{7.27}$$

in which ω_i is the mobility coefficient (the inverse of the friction coefficient). For simple diffusion the driving force is the gradient of the chemical potential μ_i. The transport equation 7.25 then becomes

$$J_i = -\omega_i c_i (d\mu_i/dx). \tag{7.28}$$

Taking the appropriate expression for the chemical potential and differentiating gives

$$\frac{d\mu_i}{dx} = \frac{d(\mu_i^\circ + RT \ln c_i)}{dx} = \frac{RT}{c_i} \frac{dc_i}{dx} \tag{7.29}$$

which, substituted in 7.28 yields Fick's law of diffusion

$$J_i = -D_i \frac{dc_i}{dx} \tag{7.30}$$

where the diffusion coefficient D_i is defined as

$$D_i = \omega_i RT. \tag{7.31}$$

If the diffusing species is an electrolyte the gradient of the *electrochemical*

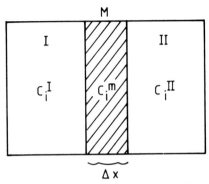

Fig. 7.3. Two compartments I and II separated by a membrane of finite thickness Δx.

potential, containing an electrical potential term, is the driving force. In this case, when the mobilities of the ions are different from each other, an electrical potential gradient is associated with the concentration gradient.

The finite thickness of membranes

In biological systems a great many of the transport processes operate by diffusion through membranes. Up to now we have considered membranes as infinitely thin barriers between cell compartments. Membranes, however, have a finite thickness and this is one of the difficulties in applying the diffusion principles, as described above, to these structures; there is no way to experimentally obtain values of the concentration gradient within the membrane phase. The solution phases bordering the membrane on each side are accessible to experimental determinations and, therefore, approximations to membrane transport processes should refer to concentration values in these phases. Figure 7.3 indicates two solution phases I and II separated by a membrane phases M, with thickness Δx. Let the concentrations of a solute i be c_i^I in phase I, c_i^{II} in phase II, and c_i^m in phase M. According to Eq. 7.28, diffusion of the solute i through the membrane phase is given by

$$J_i = -\omega_i c_i^m (\mu_i^m / dx).\tag{7.32}$$

Since the concentration gradient within the membrane phase cannot be determined and, therefore, neither can the potential gradient, an approximation in which the concentration in the membrane phase relates in a known way to the concentrations in the solution phases must be made. We can do this by using differences instead of differentials and approximating the potential gradient as

$$\frac{d\mu_i^m}{dx} \approx \frac{\Delta\mu_i}{\Delta x} = \frac{\Delta(\mu_i^\circ + RT \ln c_i)}{\Delta x}\tag{7.33}$$

in which c_i relates to the concentration in the solution phases. For small values of Δx

$$\frac{\Delta(\mu_i^\circ + RT \ln c_i)}{\Delta x} = \frac{RT}{c_i}\frac{\Delta c_i}{\Delta x}\tag{7.34}$$

With this approximation we can write

$$J_i = - \frac{\omega_i{}^m c_i{}^m RT}{c_i} \frac{\Delta c_i}{\Delta x} \tag{7.35}$$

for the flow. Of course, in general, c_i is not equal to $c_i{}^m$. We can assume, however, that at the boundaries of the membrane there exists an equilibrium concentration distribution. Thus,

$$\mu_i{}^{sol} = \mu_i{}^m \tag{7.36}$$

or

$$\mu_i{}^{osol} + RT \ln c_i{}^{sol} = \mu_i{}^{om} + RT \ln c_i{}^m \tag{7.37}$$

and, after rearranging,

$$\mu_i{}^{osol} - \mu_i{}^{om} = RT \ln (c_i{}^m / c_i{}^{sol}) \tag{7.38}$$

Since, at constant temperature, the left-hand side of Eq. 7.38 is a constant, the right-hand side must likewise be a constant. Hence,

$$c_i{}^m = k_i c_i{}^{sol} \tag{7.39}$$

in which k_i is a constant at a given temperature for a given solute. Using this in Eq. 7.35 we obtain

$$J_i = -\omega_i k_i RT(\Delta c_i / \Delta x) \tag{7.40}$$

in which we can put

$$\Delta c_i = c_i{}^{II} - c_i{}^{I}$$

Permeability
Equation (7.40) thus indicates that the flow of a solute through a thin membrane is proportional to the concentration difference across the membrane, divided by the thickness. The membrane thickness, indeed, is very small (in the order of 80 Å) and in many cases it cannot be given a precise value or meaning. Therefore, it is convenient to lump it in the proportionality coefficient and define a *permeability coefficient*

$$p_i{}^m = \frac{\omega_i k_i RT}{\Delta x} \tag{7.41}$$

for a given solute, a given temperature, and a given membrane. The flow then simply becomes

$$J_i{}^m = - p_i{}^m \Delta c_i. \tag{7.42}$$

Diffusion of ionic species through the membrane phase can be treated in the same way. The driving force in this case is the electrochemical potential gradient which contains an electrical component $z\mathscr{F}(d\psi/dx)$ as well. Making the same approximations as for the neutral solute, the flow of ion k through the membrane phase can be given by

$$J_k = -\omega_k c_k{}^m[(RT\,\Delta c_k/c_k\Delta x) + z_k\mathscr{F}(\Delta\psi/\Delta x)]. \tag{7.43}$$

Again assuming an equilibrium concentration distribution at the membrane boundaries,

$$J_k = -\omega_k k_k RT[(\Delta c_k/\Delta x) + (z_k\mathscr{F}/RT)c_k(\Delta\psi/\Delta x)]. \tag{7.44}$$

and, using the same definition for the permeability coefficient of ion k :

$$J_k = -p_k{}^m[(\Delta c_k + (z_k\mathscr{F}/RT)c_k\Delta\psi]. \tag{7.45}$$

Restrictions set by electrical neutrality prevent an electrical current from flowing across the membrane when there is no external electrical connection. If the membrane is permeable only to ion k, Eq. 7.45 can be solved for $J_k = 0$:

$$\Delta\psi = -\frac{RT}{z_k\mathscr{F}}\frac{\Delta c_k}{c_k} \tag{7.46}$$

or, after integration,

$$\psi^{II} - \psi^I = -\frac{RT}{z_k\mathscr{F}}\ln\frac{c_k^{II}}{c_k^I} \tag{7.47}$$

a result identical to that expressed in the Nernst equation 7.13 for ionic equilibrium. If the membrane were permeable to a variety of ions, with charge z_k, and each with its own permeability coefficient, $p_k{}^m$, electrical neutrality requires that

$$\sum_{k=1} z_k J_k = -\sum_{k=1} z_k p_k{}^m\,\Delta c_k - \frac{\mathscr{F}}{RT}\Delta\psi\sum_{k=1} z_k{}^2 p_k{}^m c_k = 0. \tag{7.48}$$

Solving for $\Delta\psi$, we obtain

$$\Delta\psi = \frac{RT}{\mathscr{F}}\frac{\sum_{k=1} z_k p_k{}^m\,\Delta c_k}{\sum_{k=1} z_k{}^2 p_k{}^m c_k} \tag{7.49}$$

When all ions, each with its own permeability coefficient, are taken into account this equation poses a formidable mathematical problem. Simple solutions in terms of the potential difference $(\psi^{II} - \psi^I)$ are feasible, however, because in actual biological systems the permeability of a few ions largely predominate over the permeability of other ions in the solution phases. Thus, the summation needs to be carried out only over those few ions whose permeabilities are significant in the particular membrane. The membrane potential of the resting nerve, for instance, is well approximated by Eq. 7.47

applied to K^+ indicating that the permeability coefficient of this ion dominates over that of any other ion. In the conducting stage of the nerve the permeabilities of K^+ and Na^+ change dramatically, thus providing for a potential spike which is transmitted along the nerve fiber (see Chapter 9). The permeation of a constituent, be it a neutral solute or an ion, is determined by a permeability coefficient p_i^m which contains an intrinsic mobility factor ω_i^m of the constituent in the membrane phase and a partition coefficient k_i^m. The latter factor can be seen as the relative solubility of the constituent in the membrane phase with respect to its solubility in the bordering solution phases. Both factors depend upon the constituent itself as well as the membrane, and in particular upon the mechanism, or mechanisms, by which the constituent diffuses through the membrane. Our knowledge of membrane structure is still too sketchy to be able to describe precise mechanisms. It seems probable, however, that the actual process involves the simultaneous or consecutive action of a number of mechanisms.

7.5. Transport mechanisms

Selectivity
If diffusion of a molecular species across the boundary between the solution phase and the membrane phase plays an important role, we would expect (in a medium as viscous as the lipid core of a membrane) a low mobility coefficient. For certain molecules this could be offset by a high value for the partition coefficient k_i^m. Indeed, there is a good correlation between the lipid solubility of a number of molecules and their penetration rate across cellular membranes (see Fig. 7.4). On this basis, however, one would expect hydrophylic substances, such as ions and amino acids, to be poorly permeable,

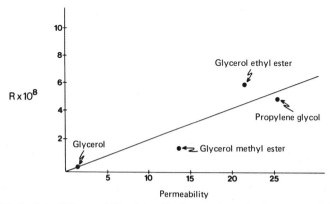

Fig. 7.4. A plot of the solubility R in oil of a number of substances and their permeability in cells of the alga *Chara*, showing the correlation between permeability and lipid solubility (data from E.J. Harris (1960) "Transport and Accumulation in Biological Systems", Acad. Press, New York).

205

whereas in many cases they are rapidly penetrating. Moreover, the high degree of selectivity of membranes with respect to different solutes cannot be accounted for by differences in lipid solubility alone.

The passing of solvent and diffusion of solutes through membranes can be visualized as taking place through channels formed by integral membrane proteins (see section 2.4). Selectivity with respect to the sign of ions could be accounted for by assuming that the channels are structured with ionic species of a given sign; the channels, thus, could act as ion exchangers.

Additional transport mechanisms have been suggested to account for the specificity of membranes. One such suggestion was induced by studies of the transport of disaccharide by bacterial membranes. These studies have led to the idea that the transport of disaccharide is effected by an enzyme called *permease*. This enzyme seems to be an induced enzyme; it is synthesized in response to the presence of its substrate. An interaction between a molecule or an ion to be transported and a substance within the membrane phase may be a more general phenomenon than the specific term permease would suggest; "carrier mechanisms" may play an important role in biological transport phenomena.

Chemical association

The specificity of enzymatic reactions is due to close noncovalent interactions between the enzyme and the substrate which lead to a substrate-enzyme complex. Thus, in biological transport systems, one could postulate a substance within the membrane phase which has a high affinity for the species to be transported. The interaction leads to an association and the resulting complex can diffuse through the membrane phase. In Fig. 7.5 such a carrier mechanism is schematically diagrammed. A substance A is bound by a carrier C, forming a complex A.C. A by itself is poorly soluble in the membrane phase so that there is very little free A present; the complex A.C can readily diffuse through the membrane. We assume further, for reasons of simplicity, that the complex is electrically neutral. If the reaction between A and C is rapid, with respect to the diffusion rate of the complex, the reaction will proceed close to equilibrium. Thus, we have the relation

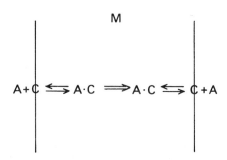

Fig. 7.5. Transport through a membrane M of a substance A by chemical association with a carrier C.

206

$$K[A \cdot C] = [A][C] = [A]([C]_0 - [A \cdot C]) \qquad (7.50)$$

in which K is the equilibrium constant and $[C]_0$ is the total concentration (bound + free) of the carrier. From 7.50 it follows that

$$[A \cdot C] = \frac{[A][C]_0}{[A] + K} \qquad (7.51)$$

To calculate the flow of the complex through the membrane phase we use Eq. 7.42 which, applied to our case, yields

$$J_{A \cdot C} = p^m_{A \cdot C} \Delta[A \cdot C] = p^m_{A \cdot C}([A \cdot C]^{II} - [A \cdot C]^{I}). \qquad (7.52)$$

If we ignore the back flow (which we could do, for instance, in an experiment in which we follow the transport of a radioactive isotope of substance A during a short time after the beginning of the experiment) we can make $[A \cdot C]^{II} = 0$. Substituting 7.51 into 7.52 we obtain

$$J_{A \cdot C} = p^m_{A \cdot C} \frac{[C]_0[A]}{[A] + K}. \qquad (7.53)$$

From 7.53 we can see that at a high concentration of A in the solution phase I the flow becomes a constant. The flow, thus, becomes saturated, just like the rate of an enzymatic reaction becomes saturated at a high concentration of the substrate. A plot of the flow as a function of the concentration of the transported substance is given in Fig. 7.6. From 7.53 it also follows that the flow is at half of its saturation value when $[A]^I = K$. This provides a means to determine, experimentally, the value K.

Saturations of the rate of transport of substances through membranes have, indeed, been measured, especially in case where there is a high degree of selectivity toward these substances. The carrier mechanism, as described above

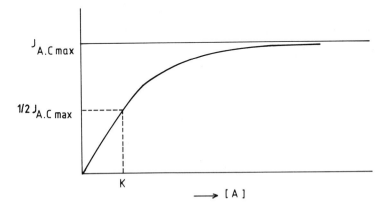

Fig. 7.6. A plot of the flow of a complex of a substance A and a carrier C through a membrane as a function of the concentration of A at one side of the membrane. At a concentration of A equal to the equilibrium constant K of the complex formation reaction, the flow is at half the saturation value.

in a simplified way, offers an explanation for the selectivity which would be difficult to imagine without some form of chemical association.

Ionophores

Some antibiotics cause changes in the permeability of specific ions. An example is the compound *valinomycin*, which shows a high specificity toward K^+ and, to a lesser extent, toward Rb^+. The antibiotic forms a complex with the ion and carries it across the membrane which is, otherwise, quite impermeable for univalent cations. Figure 7.7 shows the structure of valinomycin. Its cyclic structure enables the compound to accept the ion in its inner core, which is hydrophylic, and surround it with an outer shell, which is hydrophobic, thus permitting the ion to pass rapidly through the membrane.

Another ion-conducting antibiotic is *nigericin*. This compound catalyzes an exchange between K^+ and H^+, so that it does not cause a change in the membrane potential. Other ion conducting compounds, known as ionophores, have been shown to affect the penetration of cations across mitochondrial, chloroplast, and cytoplasmic membranes as well as across artificial membranes. Of course, the action of such compounds does not in itself imply anything about the actual transport mechanisms in biological membranes. They do make cations more soluble in the membrane phase, usually by wrapping them in a hydrophobic shell. Their specificity has to do with a more or less exact fit of the ions in their inner core. In Table 7.1 a few of these compounds are given with their different specificities for cations.

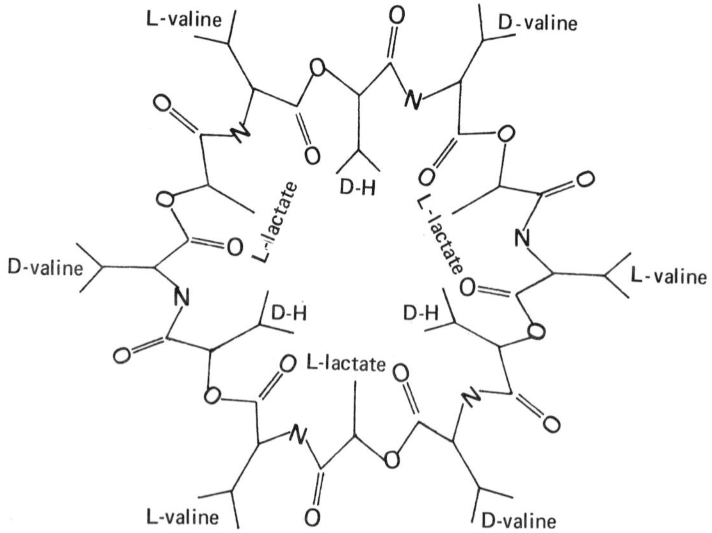

Fig. 7.7. The chemical structure of valinomycin; D-H is D-hydroxyisovalerate.

Table 7.1. Some ionophoric compounds

Compound	Ion specificity
Gramicidin A	H^+, Na^+, Li^+, K^+, Rb^+, Cs^+
Valinomycin	K^+, Rb^+
Nigericin	K^+, $-H^+$ exchange
Dinitrophenol	H^+
Carbonyl cyanide m-chlorophenyl-hydrazone (CCCP)	H^+
Fluorocarbonyl cyanide phenyl-hydrazone (FCCP)	H^+

Active transport

Many membranes translocate molecules or ions from regions of low concentration to those of high concentration. This transport against the thermodynamic gradient can be sustained only when it is coupled with an energy-supplying process. Neither simple diffusion through the membrane phase or through pores, nor the carrier mechanism *per se* can explain this translocation by active transport. It is not too difficult, however, to extend the concept of transport by chemical association in order to postulate a mechanism by which it could work. In Fig. 7.8 we have a schematic representation of a carrier mechanism similar to that of Fig. 7.6. The carrier substance C in this case, however, can be converted from a configuration of high affinity for substance A to one of low affinity and *vice versa*. This conversion can be a chemical alteration or just a conformational change. It is essentially that the conversion in one direction be coupled to an energy-yielding reaction. The complex A.C as well as the configuration of low affinity, C', can diffuse through the membrane. The process can be visualized as follows. At surface I the configuration of high affinity, C, is formed from the configuration of

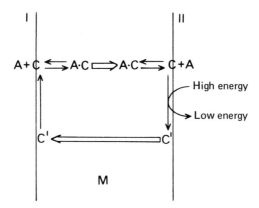

Fig. 7.8. A model for active transport by chemical association. A substance A to be transported binds to a carrier C when this carrier is in a form of high affinity for A. At each side of the membrane the carrier is converted from the high affinity form C to the low affinity form C' and *vice versa*. Both the complex A.C and the low affinity form C' can diffuse through the membrane, indicated by the open arrows. The cycle is coupled to an energy-yielding reaction.

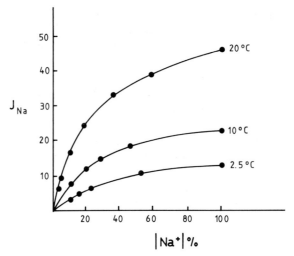

Fig. 7.9. The net flow of sodium ion J across isolated frog skin as a function of the Na^+ concentration in the bathing medium at different temperatures. The flow diminishes at lower temperatures (from F.M. Snell et al. (1965) "Biophysical Principles of Structure and Function", Addison-Wesley, Reading, MS. Reproduced with the permission of the Benjamin/Cummings Publ. Comp., Menlo Park, CA.).

low affinity, C'. Substance A then binds to C, which is continually supplied by conversion from C'. The complex A.C diffuses from I to II, where dissociation into A and C occurs. This dissociation is favored by the fact that the concentration of C is kept low because of its conversion, at surface II, to C'. C' diffuses back to surface I because its concentration at surface II is higher as a result of the continuous conversion of C into C' at surface II. The process, thus, has a cyclic character and it is evident that it cannot proceed unless it is driven by some energy-supplying reaction. Such a reaction can drive either the conversion of C into C' at surface II or the conversion of C' into C at surface I. An indication for such a mechanism is the fact that saturation curves for the net flow of ions across a number of membranes have been observed (Fig. 7.9).

The carrier C can be an ATPase and the conversion of C into C' can be a conformational change of the enzyme. One could also visualize the transport proper not occurring by diffusion but by a change of orientation of the binding sites for the substrates on the enzyme. There is some evidence for such a mechanism for the Na^+K^+ATPase. The enzyme is located in the membrane as an integral membrane protein. The following simple reaction scheme seems to give a reasonable explanation for most, if not all experimental observations: In one conformational state (E_1) the enzyme has a high affinity for Na^+ at ion binding sites located at the cytoplasmic side. In the presence of Mg^{2+} and Na^+ the enzyme binds ATP and becomes phosphorylated ($E_1.P_i$) and the ion binding sites are now facing the extracellular space. At the same time the affinity of the binding site for Na^+ decreases, leading to a release of Na^+ in the extracellular space. The affinity of the K^+ binding sites increases

210

simultaneously resulting in the binding of this ion. Then the $E_2.P_i$-form of the enzyme is hydrolyzed, resulting in the dephosphorylized E_2-form. Finally, the E_2 form is reconverted into the E_1 form transporting the K^+ to the inside of the cell.

Another ATPase which has been described in somewhat more detail is the H^+K^+ATPase found in the gastric mucosa. This enzyme plays a crucial role in the acid secretion of the stomach. It shows an extensive homology with the Na^+K^+ATPase. Also mechanistically the two enzymes have much in common. However, the Na^+K^+ATPase operates electrogenically (it creates and maintains a membrane potential by transporting three Na^+ ions to the outside and two K^+ ions to the inside in antiport for each ATP molecule hydrolyzed), while the operation of the H^+K^+ATPase is electroneutral (the hydrolysis of one ATP molecule transports two H^+ ions in antiport with two K^+ ions).

Bibliography

Bonting, S.L., and de Pont, J.J.H.H.M., (Eds.). (1981). "Membrane Transport". Elsevier, Amsterdam.

Harold, F.M. (1982). Pumps and Currents, a Biological Perspective, *Current Topics in Membranes and Transport* **16**, 485–516.

Kaback, H.R. (1974). Transport Studies in Bacterial Membrane Vesicles, *Science* **186**, 882–892.

Maunsbach, A.B. (Ed.). (1987). "NaK-ATPase". A.R. Liss, New York.

Rosen, B.P. (Ed.). (1978). "Solute Transport in Bacteria". Marcel Decker Inc., New York.

8. Membrane-bound energy transduction

8.1 The high-energy intermediate

Oxidative- and photophosphorylation

The demonstration of an ion transport-coupled ATPase system, as described in the previous chapter, suggests, of course, the possibility of reversibility. In other words, when active transport of ions can be induced by the *hydrolysis* of ATP by an ATPase, can transport of ions across a membrane in the direction of a thermodynamic gradient cause the *synthesis* of ATP by the same ATPase? An affirmative answer to this question appeared to be the solution to the problem of membrane-linked synthesis of ATP, called oxidative phosphorylation for mitochondria and photophosporylation for photosynthetic organelles.

The synthesis of ATP in the glycolytic reaction sequence (often called substrate-linked phosphorylation) occurs by enzymatic reactions in which common intermediates take care of the coupling of the energy-yielding (oxidative) and energy-conserving (phosphorylation) reactions (see Chapter 5). These reactions take place in the cytoplasmic phase of the cell. The mechanism of ATP synthesis coupled to respiratory or photosynthetic electron transport is embedded in membranes, the cristae of mitochondria, the thylakoids of chloroplasts and the inner membranes of bacteria. This linkage to an *intact* membrane system is an essential aspect of this type of energy transduction; indeed, no ATP synthesis occurs without an intact structure of the mitochondrial, chloroplast or bacterial inner membrane, although electron transport may go on unimpaired in broken membrane fragments.

The chemical hypothesis

Experimental results obtained with uncouplers (compounds which inhibit phosphorylation but leave the electron transport intact, sometimes even stimulating it) have led to the concept of the "high-energy intermediate". A hypothesis which grew up through argument by analogy from substrate-linked phosphorylation, considers this high-energy intermediate to be a chemical entity. According to this theory the coupling of the respiratory or photosynthetic oxidation-reduction reactions and the phosphorylation of ADP to ATP is accomplished as follows.

Suppose A, B, and C are consecutive components of the oxidation-reduction reaction chain. Reduction of B by AH_2 cannot proceed unless the reduced product BH_2 is complexed to an intermediate I. Thus,

$$AH_2 + B \rightarrow A + BH_2 \qquad (8.1)$$

immediately followed by

$$BH_2 + I \rightarrow BH_2.I. \qquad (8.2)$$

The reduced complex, $BH_2.I$, in turn can reduce the next component in the chain in an exergonic reaction. The free energy of this reaction is then captured in a high-energy bond designated by a "squiggle", ~, (a term coined by E.C. Slater):

$$BH_2.I + C \rightarrow B{\sim}I + CH_2 \qquad (8.3)$$

thus forming the high-energy intermediate $B{\sim}I$. In the presence of phosphate this high-energy intermediate is phosphorylated and the high-energy phosphate can in turn phosphorylate ADP :

$$B{\sim}I + P_i \rightarrow I{\sim}P + B \qquad (8.4)$$

$$I{\sim}P + ADP \rightarrow ATP + I. \qquad (8.5)$$

Uncouplers simply cause the hydrolysis of the high-energy intermediate $B{\sim}I$, thus preventing ATP from being formed while keeping the oxidation-reduction reactions running. Inhibitors of either one or both of the phosphorylation reactions, such as the antibiotic oligomycin, also stop the oxidation-reduction reactions.

A modification of the chemical hypothesis is the conformational hypothesis. In this hypothesis the high-energy intermediate is a special conformation of a factor I. This theory is represented by replacement of Eqs. 8.2–8.5 by

$$BH_2 + I + C \rightarrow B + I^* + CH_2 \qquad (8.6)$$

$$I^* + ADP + P_i \rightarrow ATP + I \qquad (8.7)$$

in which I^* designates the high-energy conformation of I.

Although the "chemical hypothesis" outlined above seemed to be compatible with any observation, it has run into difficulties. First, no one as yet has been able to demonstrate experimentally the existence of, let alone isolate, the high-energy intermediate. This by itself is no argument for rejecting the theory; the complex may be extremely unstable. More serious, however, is the fact that the theory does not account at all for the linkage to membrane structures.

8.2 The chemiosmotic model

Chemiosmotic loops
A model which rests upon a completely different kind of concept is the chemiosmotic model, proposed in its original form by P. Mitchell. It does

214

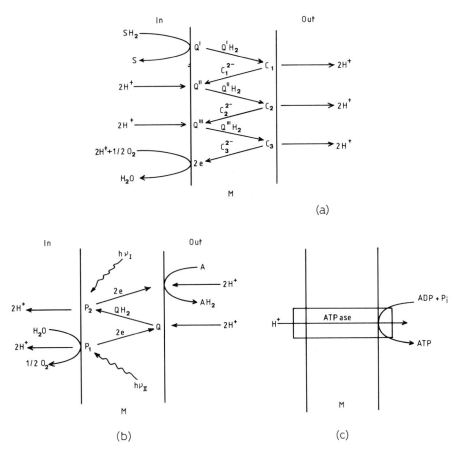

Fig. 8.1. A schematic representation of the chemiosmotic model of energy transduction. (*a*) In a mitochondrion; the oxidation of substrate SH_2 occurs by reduction of a hydrogen carrier Q which, in turn, reduces an electron carrier C at the other side of the membrane, releasing protons in the outer medium. The alternation of hydrogen and electron carriers at both sides of the membrane causes a net translocation of protons across the membrane and is essential for the hypothesis. (*b*) In a chloroplast; light causes electrons from water to flow across the membrane releasing protons in the inner medium. The reduced hydrogen carrier at the other side of the membrane picks up protons from the outer medium and donates electrons to the other light reaction. Again, the alternation of electron and hydrogen carriers at both sides of the membrane causes a net flow of protons across the membrane. (*c*) The reversed ATPase system in the membrane, synthesizing ATP at the expense of the collapse of the proton gradient.

ascribe a functional role to an intact membrane-bounded system. In fact, the high-energy intermediate in this model is a thermodynamic gradient formed across the inner membrane of the organelle (or bacterium) by the apparent translocation of hydrogen ions across the inner membrane. These hydrogen ions are translocated by an electric field which is generated by the transport of electrons from one side of the membrane to the other; the electron transport in the membrane, thus is *vectorial*. The mechanism by which this would work is illustrated schematically in Fig. 8.1a (for the respiratory chain) and 8.1b

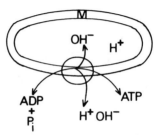

Fig. 8.2. A possible mode of operation of the reversed ATPase in the chemiosmotic hypothesis. In the reaction the components of water liberated in ATP synthesis are moved vectorially to opposite sides of the membrane by the collapsing proton gradient.

(for the photosynthetic chain).

The oxidation-reduction components (as shown in Fig. 8.1) are arranged within the membrane in such a way that electron carriers (such as the cytochromes and nonheme iron proteins) alternate with hydrogen carriers (such as the flavoproteins and quinones) at both sides of the membrane. In such a way the electrons transported across the membrane by the electron carriers, either through reduction by substrate (in the case of respiratory electron transport) or by the light-induced primary reaction (in the case of photosynthetic electron transport), generate an electric field which causes an "osmotic" translocation of protons by the hydrogen carriers either from the inside to the outside (in the case of the mitochondrial inner membrane) or from the outside to the inside (in the case of the chloroplast thylakoid). The membrane in both cases, although permeable for water, is impermeable for hydrogen ions. The translocated protons thus form a concentration gradient which, in fact, is the high-energy intermediate. An ATPase system connected to the membrane (Fig. 8.1c) can then act in reverse, and by retranslocating the protons down their gradient ATP is synthesized (see section 8.5). A simple way in which this can be visualized is shown in Fig. 8.2. The presence of the H^+ gradient in some unknown fashion, pulls the hydroxyl ion portion of the water, which is produced when ATP is formed from ADP and inorganic phosphate, to one side of the membrane and pushes the protons to the other side. In this way the net movement of the H^+ ions would produce ATP.

Uncouplers

In the chemiosmotic model, uncoupling will provide for an additional pathway for the dissipation of the proton gradient, thus preventing the ATPase from using the gradient for ATP synthesis. Indeed, all known uncouplers do have the property to "solvate" hydrogen ions in the membrane; many of them, such as dinitrophenol (DNP), carbonyl cyanide *m*-chlorophenylhydrazone (CCCP), or carbonyl cyanide *p*-trifluoromethoxypenylhydrazone (FCCP), and many fatty acids are both weak acids and lipid soluble, and are able to carry protons across a membrane (Fig. 8.3). Thus, the protonated form of the acid can enter on one side, move across to the other side, discharge its proton, and then recross the membrane as an anion. Many of the ion-carrying

216

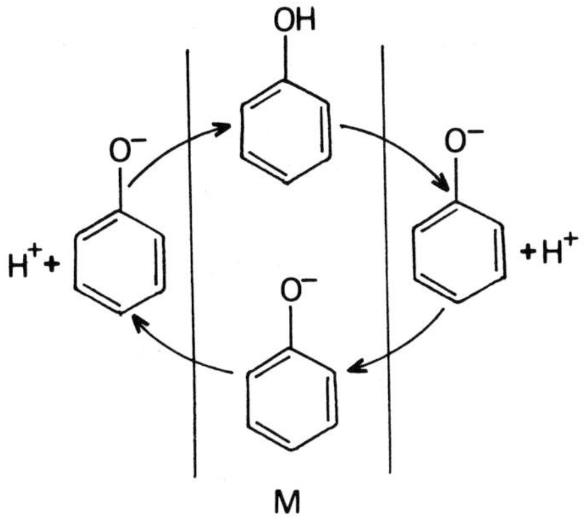

Fig. 8.3. The protonophoric action of a phenol. The phenol will carry protons in either direction as dictated by the concentration gradient. Proton transport will stop when the proton concentration is equal on both sides of the membrane.

antibiotics, or combinations of them, are uncouplers by virtue of their ionophoric properties.

Q-cycles

The chemiosmotic loops illustrated in Fig. 8.1 make explicit predictions about the redox carriers and their topology within the membrane. The electron transport chains must consist of alternating carriers for hydrogen and electrons with hydrogen carriers situated on one side of the membrane and electron carriers on the other side. Moreover, the stoichiometry in one loop cannot exeed two protons translocated for each pair of electrons passing the loop. Several of these predictions have indeed been confirmed. For example, many dehydrogenases are found on the matrix (or stroma) side of the membranes, while cytochromes of the *c*-type are exposed to the internal space of the vesicles. However, some later findings appeared to be inconsistent with the original hypothesis. In particular, the role and the place of several *b*-type cytochromes were difficult to be incorporated in the model. In mitochondria, two *b*-type cytochromes carry electrons across the membrane but a *c*-type cytochrome promotes the *reduction* of these cytochromes, an effect which is enhanced by the potent inhibitor of electron transport in the b/c_1 complex, antimycin A.

In order to account for such difficulties a modification of the chemiosmotic loop was devised (by Mitchell himself). This modification became known as the *Q-cycle*. Several forms of Q-cycles, adapted to additional observations on electron transport, proton translocation and stoichiometry, in mitochondria, chloroplast thylakoids, and chromatophores of photosynthetic bacteria were developed since. A simplified version of a Q-cycle in a photosynthetic bacterium is illustrated in Figure 8.4. Quinone (in bacteria ubiquinone) plays the central role. It is reduced to a semiquinone anion by an electron coming from the reaction center, after which it is protonated by a proton from the outside of the chromatophore. The semiquinone is next reduced to quinol by the

217

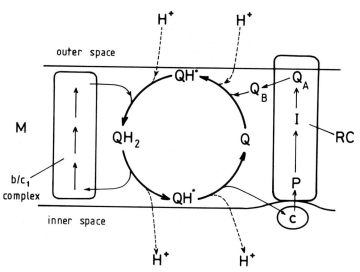

Fig. 8.4. A simplified version of a Q-cycle in the membrane M of a photosynthetic bacterium. RC is the reaction center; Q_A and Q_B are the primary and secondary acceptor; Q is quinone; QH· is semiquinone radical; QH_2 is quinol; c is cytochrome c. Explanation in the text.

uptake of a second proton and an electron coming from a *b*-type cytochrome (possibly a part of the b/c_1 complex). The quinol is reoxidized, firstly to semiquinone by returning an electron to the b/c_1 complex, releasing a proton to the inner space and subsequently to quinone by donating an electron to cytochrome c and releasing another proton into the inner space. Electron transports within the b/c_1 complex and from cytochrome c to the reaction center then close the cycle.

8.3 Proton translocation

Proton motive force
In chloroplast membranes a reversible uptake of protons during light-induced electron transport has been detected and is now a common observation. Such a light-induced proton uptake can also be demonstrated in membrane-bounded vesicles (chromatophores) derived from photosynthetic bacteria, although substantially lesser in extent. Mitochondria exhibit an extrusion of H^+ but submitochondrial particles take up protons when supplied with respiratory chain substrates (NADH, or succinate). This is a manifestation of the *sidedness* of the membrane; the disruption of the inner membrane, when intact mitochondria are subjected to high pressure or ultrasonic vibration, is predominantly at the "neck" portions of the sharply folded cristae (see Fig. 8.5). The sub-mitochondrial particles, thus are "iside out" as compared to intact mitochondria. This also is the way in which chromatophores are formed from

218

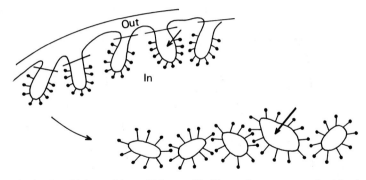

Fig. 8.5. Submitochondrial particles which are "inside out" as compared with the intact mitochondrion. Upon disruption of the organelle the inner membrane shears preferentially in the "necks" of the inward folded cristae. The ATPsyntase complexes (the knobs on stalks), which in the intact mitochondrion pointed inward, are sticking to the outside in the submitochondrial particles.

photosynthetic bacteria.

Redox reactions, thus, generate an electrochemical gradient of protons given by

$$\Delta\tilde{\mu}_{H^+} = \tilde{\mu}_2 - \tilde{\mu}_1 = RT \ln ([H^+]_2/[H^+]_1) + \mathscr{F}(\psi_2 - \psi_1). \qquad (8.8)$$

Since pH= $-\log$ [H$^+$],

$$-\Delta pH = -(pH_{(2)} - pH_{(1)}) = 2.3 \ln ([H^+]_2/[H^+]_1),$$

and

$$\Delta\tilde{\mu}_{H^+} = \mathscr{F} \Delta\psi - 2.3 \, RT \, \Delta pH \qquad (8.9)$$

in which $\Delta\psi = (\psi_2 - \psi_1)$. The electrochemical potential can also be expressed in the electrical unit, Volt; Eq. 8.9 then becomes

$$pmf = \frac{\Delta\tilde{\mu}_{H^+}}{\mathscr{F}} = \Delta\psi - 2.3 \frac{RT}{\mathscr{F}} \Delta pH. \qquad (8.10)$$

The quantity pmf is called the *proton motive force*. From 8.10 it can be seen that the pmf consists, in fact, of two components. The electrical component, $\Delta\psi$ is the membrane potential; the chemical (or osmotic) component is determined by the pH gradient, ΔpH.

Counterion flow

According to the chemiosmotic model, energization of the membrane (in photosynthetic membranes by light, in respiratory membranes by substrate oxidation) initiates a net flow of protons across the membrane and, at the same time, a "charge generating" transfer of electrons across the membrane in the opposite direction. Since the electric capacity of membranes is relatively small (typically 1 μF/cm^2 in a very small membrane vesicle), a small number

219

of charges translocated is sufficient to generate large membrane potentials in a very short time (usually small fractions of a second). This rapidly formed electric potential is transient, however; it is diminished by slower secondary electrophoretic counterion fluxes. As a result of these ion fluxes, the ΔpH builds up more slowly. Thus, the electric component of the pmf is gradually replaced by a pH gradient; the two components are formed out of phase.

Counterion movements are indeed demonstrated in chloroplast preparations. In particular, a symport movement (movement in the same direction) of chloride ions simultaneous with proton uptake is conspicuous. Antiport movement (movement in the opposite direction) of K^+ and Mg^{2+} simultaneous with proton uptake has also been detected but the latter does not seem to be the important compensating counterion flow in these organelles. Antiport movement of K^+ and Na^+ seems to be the membrane potential compensating process in mitochondria.

The situation in photosynthetic bacteria, at least in chromatophore preparations, is somewhat different; the bacterial membrane apparently shows a greater resistance to penetration of monovalent ions. K^+ and Na^+ have only low permeability values; the same holds for Cl^-. This is why proton uptake in chromatophores from photosynthetic bacteria is smaller than that in chloroplast thylakoids.

8.4 pH gradient and membrane potential

The measurement of ΔpH

It is important to have quantitative data about the two components of the proton motive force. Unfortunately, none of the techniques presently available give unique and undisputed results; they are often subject to certain assumptions which are very hard to test. At best, estimates can be made which, nevertheless, are valuable for the qualitative evaluation of proposed mechanisms.

The determination of the ΔpH particularly is difficult. Steady-state values of this component can be estimated by using techniques which are based on equilibrium distributions of radioactive or fluorescent amines and of NH_4^+. An example is the fluorescent dye 9-amino acridine. This dye can bind protons as a result of which it becomes charged. Assuming that only the uncharged forms of the dye permeates the membrane and that the fluorescence is quenched when it is inside the vesicle, the fluorescence yield is a measure of the distribution of the protons inside and outside.

Kinetic measurements cannot be made with such techniques because of the slow response of these dyes. There are, however, pH-indicating dyes that respond to changes of the pH in the millisecond range or even faster by changing their absorption spectrum. Binding of such dyes to the membranes would be a problem but does not appear to be significant for dyes such as chlorophenol red, phenol red, cresol red and phenol violet. Rapid changes of the *external* pH can be followed that way. Rapid measurements of the *internal* pH of membrane vesicles is more problematic, however. Some people claim that the dye *neutral red* behaves as a true pH indicator for transients occurring in the internal phase of membrane vesicles if the pH transients outside the vesicle are selectively buffered away by nonpermeant bovine serum albumin (see, for instance, W. Junge and J.B. Jackson, Chapter 13 of "Photosynthesis", Vol. 1, mentioned in the Bibliography).

The measurement of $\Delta\psi$

The membrane potential in steady-state conditions can be measured by determining the equilibrium distribution of a permeable (radioactive) ion. The Nernst equation (Eq. 7.13) then gives the value of the (bulk phase) membrane potential. This technique, however, is based on rather rough approximations; the volume of the internal compartment is at least three orders of magnitude smaller than that of the external medium, so that the internal concentrations can reach very high values. Therefore, the assumption of the activity coefficient being one may not be valid. In fact, the possibility of ion binding to the membrane or other kinds of complex formation, which may alter the distribution ratio, is completely ignored.

Another method to measure membrane potentials is based on spectroscopic effects of the electric field on intrinsic or added pigments. Obviously, this method is used most widely on photosynthetic membrane vesicles, which are loaded with pigment molecules. A typical value of the membrane potential is about 200 mV. In a membrane which is about 80 Å thick, this results in an electric field of 2.5×10^{10} V/m. This, clearly, should have, and has, an effect on the spectroscopic properties of the intrinsic pigment molecules. Such effects are known as *electrochromic* effects.

Electrochromic shifts

Electrochromic effects (such as a Stark-effect, for example) result in changes of the absorption spectra, in particular changes in the extinction, splitting of absorption bands and red and/or blue shifts of absorption bands. Only the latter type of absorption change seems to be dominant in the membranes of photosynthetic organisms. These changes show up in particular in the spectral region where carotenoids absorb. Carotenoid spectra have a typical three-banded structure in the region between 400 and 500 nm. In photosynthetic bacteria, this region is conveniently separated from the regions in which chlorophyll and other pigments absorb and, therefore, the spectra are easily recognizable.

If a molecule has an electric dipole μ°, the total dipole moment in an electric field E would be

$$\mu^E = \mu^\circ + \alpha.E \qquad (8.11)$$

in which α is the polarizability tensor. The energy of the interaction with the field is

$$U^E = -\int \mu^E.dE. \qquad (8.12)$$

Hence, the energy of a dipolar system in an electric field is

$$U^E = U^\circ - (\mu^\circ + 1/2\ \alpha.E).E \tag{8.13}$$

in which U^E and U° are the energies in the presence and in the absence of an electric field respectively. This is true for the ground state as well as for the excited state. If the ground state and the excited state differ in dipolar character, the energy difference between the states changes. This change is reflected in a frequency shift

$$h\Delta v = - [(\mu^\circ_e - \mu^\circ_g) + 1/2(\alpha_e - \alpha_g)E].E \tag{8.14}$$

in which the subscripts e and g refer to the excited and ground states respectively, Δv is the frequency shift and h is Planck's constant. The wavelength shift then is given by

$$\Delta\lambda = \frac{\lambda^2}{hc}(\Delta\mu + \Delta\alpha.E).E. \tag{8.15}$$

In chromatophores from photosynthetic bacteria, light induces a red shift of the three carotenoid absorption bands (Fig. 8.6a). This shift is due to the electrical potential created by the electron transport as well as by the membrane potential component of the proton gradient. The electrical origin of the shift can be proved by an experiment in which a solution of KCl is injected, in the dark, in a suspension of chromatophores containing valinomycin. This "pulsed" addition causes a transient change of the absorption in the carotenoid absorption spectral region (see Fig. 8.6b). After a while, diffusion of counterions will neutralize the potential and as a result the concentrations of K^+ inside and outside will become equal. Another "pulsed" addition of KCl will again cause a transient absorption change. The spectrum of the transients, as shown in Fig. 8.6c, is identical to the difference spectrum of the light-induced carotenoid absorption band shifts.

Assuming that after equilibration the outside K^+ concentration is equal to the inside K^+ concentration, one can use this technique to calibrate the absorption band shift in terms of a membrane potential in volts by applying the Nernst equation (Eq. 7.13). Doing this, one finds that the magnitude of the transient carotenoid absorption change induced by the addition of KCl is a *linear* function of the logarithm of the outside concentration of K^+, hence of the membrane potential, as shown in Fig. 8.7. This is surprising; carotenoids, since they are symmetric molecules, do not have a permanent dipole moment. Hence, $\Delta\mu = 0$ and

$$\Delta\lambda = \frac{\lambda^2}{hc}\ 1/2\ \Delta\alpha\ E^2. \tag{8.16}$$

The wavelength shift thus should be a *quadratic* function of the electric field. This seeming contradiction between experiment and theory could be solved when it was demonstrated that polar chlorophyll molecules close to the carotenoids, induce dipoles in the latter; the linearity thus appeared to be

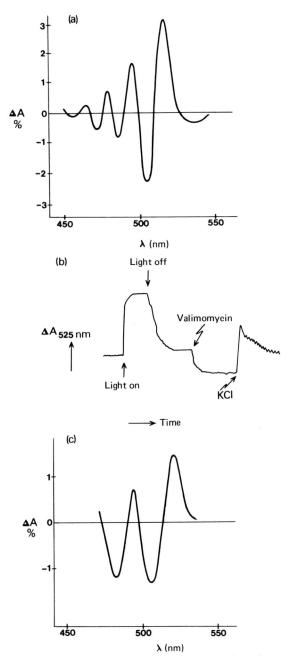

Fig. 8.6 (*a*) A light-minus-dark difference spectrum of a suspension of chromatophores from the purple bacterium *Rhodobacter sphaeroides*. The spectrum shows the light-induced red shift of the carotenoid absorption bands. (*b*) A trace of the changes of the absorption of the chromatophore suspension at 525 nm. Light induces a reversible increase of the absorption; addition of valinomycin causes a small decrease, and subsequent addition of KC1 causes a transient increase of the absorption. (*c*) The spectrum of the transient change in absorption upon addition of KC1 in the presence of valinomycin to the chromatophore suspension. The spectrum shows clear similarities with the light-induces difference spectrum presented under (*a*).

223

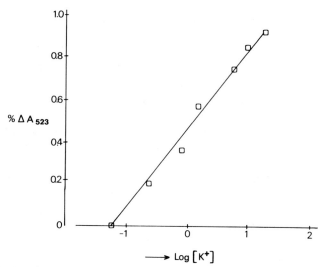

Fig. 8.7. A plot of the transient increase of the absorption of a suspension of chromatophores from *R. sphaeroides* at 523 nm as a function of the logarithm of the final outside concentration of K^+ in the presence of valinomycin.

a *pseudolinearity* since the light-induced electric field is at least an order of magnitude smaller than the induced dipole field.

Electrochromic absorption changes have also been measured in algae and chloroplasts. A rapid light-induced increase of the absorption in the spectral range between 515 and 530 nm accompanied by a decrease in the 495 nm region can be attributed to electrochromism; this absorption change has been used to estimate transient membrane potentials. In mitchondrial membranes there are no intrinsic pigments like in photosynthetic systems. Estimates of the membrane potential in these systems have been made by incubating extrinsic pigments, such as cyanine dyes or oxonols in the membrane. However, these probes do not behave at all like the intrinsic probes of the photosynthetic systems and there is still controversy about their mechanisms.

ΔpH- and Δψ-driven ATP synthesis
Light-induced spike potentials (due to electron transport across the membrane) of 420 mV and light-induced steady-state potentials (due to a steady-state equilibrium between proton translocation and counterion fluxes) as high as 240 mV have been determined in chromatophores from certain species of photosynthetic bacteria. In spite of the disputes that still go on about the quantitative aspects of the determination, these values demonstrate that the electrical component of the pmf is substantial, at least in some species of photosynthetic bacteria. Using the same calibration technique, it can be shown that in chloroplasts, although a substantial transient change (upon illumination with a short flash of light) can be detected, the "steady-state" membrane potential measured in continuous light is at best some 10 mV. This, undoubtedly, is due to a more extended counterion flow, probably mostly symport movement of Cl^-. Thus, in chloroplast under steady-state conditions, the electrical

224

component is very small and the pmf is largely in the form of a pH-gradient.

The effect of the ionophoric uncouplers on the membranes of different organisms illustrates clearly this difference. We have seen that these uncouplers do move ions across membranes. Nigericin, for example, exchanges H^+ ions for K^+ ions. Addition of nigericin to a suspension of electron transporting membrane bounded vesicles, thus abolishes the pH gradient but leaves the membrane potential, if there is any, intact. The result of such an experiment is that, in mitochondrial preparations as well as in chloroplast thylakoids, ATP synthesis is uncoupled but in bacterial preparations it is not. In the latter systems the membrane potential is sufficient to drive a proton flux through the ATPsynthase, while in chloroplasts and mitochondria no membrane potential is left to do so, after the pH gradient has been destroyed. Valinomycin, another ionophore, moves K^+ ions across. The effect of this antibiotic, therefore, is to destroy the membrane potential, again if there is any, leaving the pH gradient intact. Valinomycin, in the presence of K^+, does not uncouple ATP synthesis in any of the three types of membrane bounded vesicles unless a pH gradient-destroying agent, such as nigericin, is also present. In bacterial preparations nigericin also works as an uncoupler when a permeant anion (such as the anion of thiocyanate, CNS, which easily permeates through a bacterial membrane) is present. Valinomycin does, however, cause an appreciable time lag in the onset of ATP synthesis in chloroplasts, which is in agreement with the concept that the electrical component of the pmf forms very rapidly in photosynthetic membranes, much more so than the chemical component and that ATP synthesis is very sensitive to the destruction of the membrane potential.

In chromatophores of photosynthetic bacteria, the transient potential rise is due to, what is called *electrogenic* electron transport steps. Three distinct phases have been established. These phases are linked to, succesively, the primary electron transport in the reaction center, the rereduction of the reaction center bacteriochlorophyll by cyrtochrome c, and secondary electron transport in the b/c_1 complex. In chloroplasts the primary electron transport of the two photosystems have been shown to contribute equally to the transient rise of the potential. A slower secondary rise is associated with secondary electron transport.

8.5 The proton-translocating ATPase

ATPase activity
As has been suggested in the first section of this chapter, the ATPase system acting in the reverse by synthesizing ATP can also *hydrolyse* ATP. When it does so, protons are translocated in the opposite direction. Thus, also hydrolysis of ATP can generate a proton electrochemical gradient. One can truly say that this remarkable enzyme complex is a proton-translocating ATPase with a reversible function. It is a universal feature of energy-transducing membranes. It is present in mitochondria, chloroplasts, aerobic and photosynthetic bacteria

and even in bacteria which do not have a respiratory system, relying exclusively on fermentation. The structure of the complex is very similar in all these cases. It operates either as a synthase, utilizing $\Delta\tilde{\mu}_{H^+}$ for the synthesis of ATP or, as in the case of the fermenting bacteria, utilizing ATP to generate $\Delta\tilde{\mu}_{H^+}$ for the purpose of transport.

Structure

Originally called the coupling factor, the proton-translocating ATPase can be recognized under the electron microscope as the mushroom-like knobs projecting from one side of the membrane (see Figure 5.14, Chapter 5). In the intact organelle or cell the knobs always project into the inner space, in mitochondria into the matrix, in chloroplasts into the stroma and in bacteria into the inner cytoplasmic space. In chromatophores and sub-mitochondrial particles the membranes are turned inside-out so that here the knobs project outward. This orientation corresponds to the function; ATP is synthesized or hydrolysed on the side of the membrane from which the knobs project, while protons cross the membrane, during ATP synthesis from the side without the knobs and during hydrolysis from the side with the knobs.

The complex can be seen as consisting of two parts. The part that, under the electron microscope, is seen as the knobs projecting from the membrane, is called F_1 (in mitochondria) or CF_1 (in chloroplasts). It contains the catalytic site for ATP synthesis or hydrolysis. The other part is called F_0 (or CF_0). It consists of hydrophopic polypeptides and is buried inside the membrane. It appears to be responsible for the conduction of protons through the membrane. F_1 (or CF_1) can be detached from F_0 (or CF_0), and, hence, from the membrane by a variety of treatments. Both, the separated F_1 and CF_1 are soluble and have molecular weights of about 350,000. They contain at least five subunits, termed α, β, γ, δ, and ϵ, some of them present in multiple copies. Although the subunit composition is remarkably similar in preparations from different sources, there is no consensus yet about the exact number of

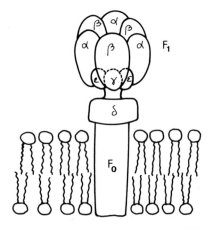

Fig. 8.8. A schematic representation of the protontranslocating ATPase in a membrane.

copies of each subunit in the complex. Even less is known about the structure of F_0 (or CF_0). It probably has multiple copies of three types of subunit in addition to some proteins that bind inhibitors (see below). A very preliminary and schematic representation of the complex as it is anchored in, and sits on the membrane is given in Fig. 8.8.

Function
Soluble F_1 (or CF_1) catalyses a rapid hydrolysis of ATP but the reverse reaction, the synthesis of ATP has never been observed in any of the preparations. The complex looses gradually its catalytic activity when it is dissected further. Experiments of this type have given some evidence that the catalytic site is associated with the β-subunit. The δ-subunit appears to be required for the binding of F_1 to F_0. Little is known about the other subunits.

In mitochondria and in sub-mitochondrial particles, the function of the complex can be inhibited by the antibiotic *oligomycin*. Unlike uncouplers, this inhibitor stops ATP synthesis as well as electron transport. It binds to a protein that appears to be required for the binding of F_1 to F_0. If olygomycin is bound to that protein, it interferes with the movement of protons through F_0. The protein, therefore is called "Oligomycin-Sensitive-Conferring-Protein" or OSCP. Bacteria and chloroplasts lack this protein and, therefore, oligomycin has no effect in these systems. Another protein, present in all energy-transducing membranes, binds the compound *dicyclocarbodiimide* (known by its acronym DCCD) with an effect similar to that of oligomycin in mitochondrial systems. DCCD-binding proteins appears to be part of the F_0, respectively CF_0 complex.

The inhibitory action of oligomycin and DCCD has given evidence for the function of F_0, respectively CF_0 as proton conductors. When, for example, F_1 is detached from submitochondrial particles, the resulting F_1-deficient particles not only stop synthesizing or hydrolyzing ATP but are also inefficient in the other energy-dependent processes. In fact, the removal of F_1 has the same effect as adding an uncoupler. The addition of oligomycin restores the energy-transducing processes, as does the rebinding of F_1. Normally, proton conduction is linked to ATP synthesis (or hydrolysis). Removal of F_1 leaves an uncontrolable proton flow, which, however, is still inhibitable by oligomycin or DCCD.

Mechanism of $\Delta\tilde{\mu}_{H^+}$-driven ATP synthesis
It is not known how the ATPsynthase uses the proton electrochemical gradient to make ATP. One could think of a direct mechanism, such as the one that was originally proposed by Mitchell. In such a model the protons driven by $\Delta\tilde{\mu}_{H^+}$ react with the oxygen of inorganic phosphate forming water and leaving a reactive species that reacts directly with ADP to form ATP. An alternative mechanism in which the translocated protons play a less direct role is, however, also possible. In such a model the free energy of the proton gradient is used to induce conformational changes in the ATPsynthase. The conformational changes, in turn, would cause the production of ATP by changing the affinities of F_1 for the substrates and the products of the reaction. The actual formation

of ATP at the catalytic sites would involve relatively small energy changes, whereas the release of ATP from F_1 would be the major energy-requiring step. Such a mechanism would be based on the same principle as that of muscle contraction, for which there is much more evidence. In muscle (see Chapter 10) the large change of free energy is not exhibited by the hydrolysis of ATP but by the release of phosphate (and ADP) from the actomyosin complex.

There is some evidence to support such a model; conformational changes in the ATPsynthase upon energization have been shown to occur and both F_1 and CF_1 possess tight binding sites for adenine nucleotides. A conformational change, however, does not by itself explain the coupling of ATP synthesis to proton translocation. At the time of this writing, this problem still is a central issue in bioenergetical research.

Bibliography

Fillingame, R.H. (1980). The Proton-Translocating Pumps of Oxidative Phosphorylation, *Ann Rev. Biochem.* **49**, 1079–1113.

Mitchell, P. (1961). *Op. cit.* (Chapter 5).

Nicholls, D.G. (1982). "Bioenergetics; An Introduction to the Chemiosmotic Theory". Acad. Press, London, New York.

Racker, E. (1976). *Op. cit.* (Chapter 5).

Schuldiner, S., Rottenberg, H., and Avron, M. (1972). Determination of ΔpH in Chloroplasts; Fluorescent Amine as a Probe for the Determination of ΔpH in Chloroplasts, *Eur. J. Biochem.* **25**, 64–70.

Slater, E.C. (1983). The Q cycle, an Ubiquitous Mechanism of Electron Transfer, *TIBS* **8**, 239–242.

9. Biophysics of nerves

9.1 Nerves

Exitability
A characteristic property of living organisms is excitability, that is the ability to react upon a stimulus from without. Changes in the immediate environment of the organism (a stimulus) evoke specific changes in the organism itself or in some parts of it (a response). A clear example is phototaxis, the light-triggered movement of photosynthetic microorganisms (see Chapter 10). Another example is the contraction of a muscle fiber upon an electric stimulus. In all these cases the response is rapid (in the order of milliseconds). Although an organism can also have relatively slow responses to changes in its environment (for instance the induced synthesis of enzyme in response to substrate concentration, which takes minutes or even hours), the term excitability is reversed for the fast responses. All such phenomena seem to be closely associated with the distribution of charge across membranes, particularly the cytoplasmic membrane.

Neurons
A cell in which excitability is particularly developed is the nerve cell. A nerve cell is called a *neuron* (Fig. 9.1). It consists of a cell body from which protrude small processes called *dendrites* and one large process called an *axon*. Many of the larger axons are "wrapped" in a sheath, called *myelin sheath*, consisting of several layers of lipid membrane (see Chapter 2). The myelin sheath is

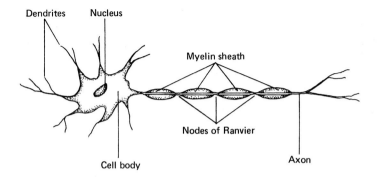

Fig. 9.1. A drawing of a neuron, showing the cell body, the myelin sheath covered axon, and the dendrites.

229

periodically interrupted, forming the nodes of Ranvier. Some axons can be very long; for instance, the neurons which control the muscles in the human finger have their cell bodies in the spinal cord.

Clusters of neuron cell bodies are called *ganglia* (singular, *ganglion*). Bundles of axons are called *nerves*. The basic unit structure of a nerve is the *funiculus*, a bundle of axons covered with connective tissue and supplied with blood vessels. A bundle of funiculi, covered by a second sheath of connective tissue makes up the nerve.

Connections between neurons are called *synapses* (see section 9.3). They occur between branched ends of axons, dendrites, and collateral branches of axons of different neurons. Axons can also end in a synapse at receptors, such as those of the sense organs, or, like in the motor nerves, at the end of a muscle fiber. The endplate of a muscle fiber is a special structure at which an action potential is generated by a synaptic stimulation from an axon of a motor nerve.

9.2 The action potential

Membrane potential

When metabolism is maintained in a cell a characteristic potential difference exists across the cytoplasmic membrane. This potential difference can vary between 50 and 100 mV and is positive on the outside. By careful insertion of a microelectrode the potential difference of many types of cells can be measured and compared to the concentration distribution of permeant ions. The latter, as we have discussed in Chapter 7, must satisfy the Nernst equation (Eq. 7.13). Most cytoplasmic membranes have relatively large permeabilities for K^+ and Cl^- and a much smaller permeability for Na^+. The membrane is virtually impermeable to other ions.

The origin of this membrane potential is the very distribution of the K^+ and Na^+ cations. This distribution is maintained by active transport of the cations similar in form to the ATPase activity described earlier. K^+ and Na^+ are actively transported in antiport, K^+ to the inside Na^+ to the outside, by one ATPase. In the absence of K^+ from the external medium no Na^+ extrusion can be detected. Since the antiport transport causes movement of equal charges in both directions, the active transport by itself cannot result in a membrane potential. However, the passive leak of K^+ to the outside is much more rapid than the leak of Na^+ to the inside, owing to the differences in permeability of the two ions. Thus, there is a net movement of positive charge to the outside which builds up a positive potential at the outside sufficiently great to oppose further leakage of K^+. At this point there is a steady state maintained by the activity of the Na^+K^+ATPase and the membrane potential is given by the Nernst equation applied to K^+.

Action potential

The membrane potential described above is called the *resting potential*. A disturbance of the membrane, in other words a stimulus (either mechanical, chemical, or electrical), upsets this balance and causes a transient change in the membrane potential, called an *action potential*. The action potential does not occur simultaneously over the entire membrane surface but is localized at the spot where the disturbance (the stimulus) took place. However, the potential, after its initiation, moves as a wave over the membrane surface away from the spot. In the case of an elongated fiber, such as the axon of a nerve cell (Fig. 7.1) the action potential passes along the fiber at a constant rate, thus constituting the nerve impulse.

Although action potentials are an essential feature of the function of nerve and muscle cells (they are used to transmit messages in these cells), they are by no means confined to only this type of cell. Action potentials can be evoked in virtually every type of cytoplasmic membrane; they can be considered as a means of rapid communication between different regions of the cell in order for the cell to respond as a whole to a local stimulus. However, the phenomenon is best studied in nerve and muscle cells and we shall discuss it as it occurs in these cells.

A single axon can be removed from an organism and, for a certain period, be examined in the laboratory. Moreover, one can uncover a nerve cell, or part of the nervous system, in a living animal and test it under *in vivo* conditions. In all such experiments it is relatively easy to stimulate the nerve with an electrical stimulus and observe the effect by electrodes placed in or on the cell membrane. A number of properties of the action potential can be demonstrated by this kind of experimentation:

1. The action potential is a potential spike with an amplitude that does not depend upon the amplitude of the stimulus. It has a typical value of about 130 mV and occurs in a regenerative way when the stimulus reaches a certain threshold value (Fig. 9.2). The spike is an all-or-none effect, suitable for conveying messages in a binary code (on-or-off, 1 or 0); its duration is about a millisecond.

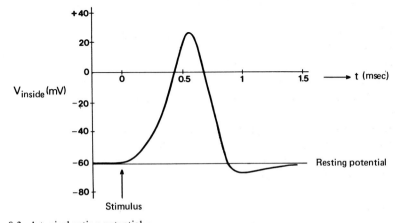

Fig. 9.2. A typical action potential.

231

2. The spike has a polarity which is opposite to that of the resting potential. Thus, it is negative at the outside, positive at the inside.
3. There is a refractory period immediately after the passage of the spike of about 1 msec during which a stimulus does not evoke an action potential, no matter how strong it is. Because of this feature no mixing of spikes can occur and the frequency of occurence is limited to less than 1000/sec.
4. The potential spike evokes new spikes in adjacent regions of the membrane. Because of the refractory period, this effect is as if the spike were propagated in both directions from the point of stimulation, like a traveling wave. Since the ocurrence of each action potential is regenerative, there is no attenuation of the "wave" as it travels along the membrane. The velocity at which the spike moves varies between 50 and 150 m/sec in vertebrates.

Na⁺ and K⁺ permeabilities

Na^+ and K^+ permeabilities

It seems obvious that the action potential, which is a transient departure from the resting potential, is a transient change of the steady-state balance of the ion concentration distribution across the cytoplasmic membrane. This, indeed, appears to be true. In fact, what happens is (see Fig. 9.3) that when a stimulus exceeds a threshold value, the permeability for Na^+, which is low in the resting state, increases abruptly and Na^+ floods into the cell. The membrane potential becomes more positive inside the cell which again increases the Na^+ permeability, making the event explosive. The potential even becomes depolarized (positive at the inside) until it reaches a maximum, after which the permeability for Na^+ decreases again. Meanwhile K^+ leaks out and this tends to restore the original membrane potential. The final phase of the action potential is the operation of the K^+Na^+ATPase to exchange the cations to their original concentration levels. Thus, due to the increase of the Na^+ permeability, a Na^+ current starts to flow in, followed by a K^+ current to the outside. This causes the transient depolarization of the membrane potential.

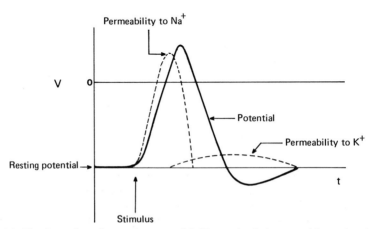

Fig. 9.3. The formation of an action potential. Upon stimulation a rapid transient increase of the permeability to Na^+ occurs which is followed by a slower increase of the K^+ permeability. As a result Na^+ will flow in and K^+ will flow out, thus creating the regenerative potential spike.

232

When the Na⁺ permeability returns to the resting value, the K⁺ current restores the potential to a level slightly below the resting potential. The resting state is then restored by the $K^+Na^+ATPase$ activity. During this process the membrane is refractory to new stimuli.

The transient inward Na⁺ current produces charge dislocations in adjacent regions, causing an increase in Na⁺ permeability. The sequence of events is thus repeated in adjacent parts of the membrane and the action potential is propagated away from the point of stimulation. Because of the refractory period the action potential never will move back.

This picture of spike generation in nerve and muscle cells has emerged from meticulously carried out experiments in which action potentials are evoked, measured, and controlled by microelectrodes. Ionic fluxes across membranes of nerves have also been measured, using radioactive isotopes of Na and K. The results of such experiments are fully consistent with the mechanism of action potential generation described above.

Voltage clamp
The events leading to the development of an action potential are too rapid to be followed by the current methods for the assay of ion concentrations. A type of experimentation that circumvents this problem is the *voltage clamp*. In these experiments the membrane potential is clamped at a given value by a feedback system which introduces a counter current opposing changes in the membrane potential from the preset value (see Fig. 9.4). When the membrane in such an experiment is clamped at a voltage exceeding the threshold value for initiating the action potential, a transient current flows inward for a few milliseconds and then the curent reverses and flows outward. The transient

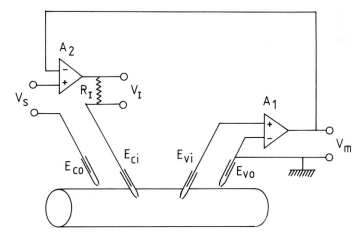

Fig. 9.4. An illustration of the principle of the voltage clamp. The membrane potential is measured by the microelectrodes E_{v_i} and E_0 which are connected to an amplifier A_1 with a gain of one. The output of this amplifier is fed back to an open loop operational amplifier A_2, which supplies the current necessary to make the membrane potential equal to a preset potential V_s. The current is fed through the membrane by the current electrodes E_{c_i} and E_0. In this way the membrane potential is held on the preset value V_s.

233

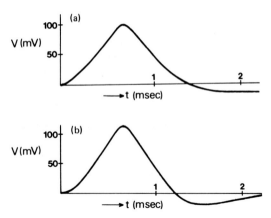

Fig. 9.5. A comparison between an action potential calculated from the Hodgkin-Huxley equations (*a*) and a measured action potential (*b*).

inward current was found to be a "sodium current" while the outward current replacing it after a short time was a "potassium current". Of course, no action potential can develop in such an experiment because the voltage is clamped at the preset value. This behavior has been empirically analyzed and a set of differential equations has been derived which are generally known as the Hodgkin-Huxley equations. These equations can be used to characterize the behavior of the nerve fiber (the axon) in terms of changes in the permeability of the membrane to Na^+ and K^+. Numerous checks have been carried out and the equations, as well as the molecular model of action potential generation based upon transient changes in permeability to Na^+ and K^+ have been found to be essentially correct (Fig. 9.5). One could also fit the transmission of action potentials between different cells in the synapses by means of chemical transmitters into these equations.

9.3 Synapses

The synaptic event
We have stated above that the propagation of an action potential along a nerve axon occurs in both directions from the point of the stimulus. This would not make much sense in a nervous system which is meant to carry messages from one point to another unless there were some kind of rectifying system. Such rectifying systems are present at the *synapses*, junctions of one (nerve) cell and another (illustrated in Fig. 9.6). The terminal end of the presynaptic fiber (usually, but not always, an axon), is separated from a receptor surface of the postsynaptic fiber (usually a dendrite, an endplate of a muscle fiber, or a secretory cell of the endocrine system) by a small gap (about 1 μm) called the *synaptic cleft*. When an action potential arrives at the terminal of the presynaptic fiber, it triggers the release of chemical substance, called

234

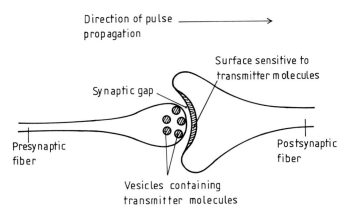

Direction of pulse propagation

Surface sensitive to transmitter molecules

Synaptic gap

Presynaptic fiber

Postsynaptic fiber

Vesicles containing transmitter molecules

Fig. 9.6. A synapse. When an action potential arrives at the terminal end of the presynaptic fiber, synaptic transmitter is released from the synaptic vesicles containing neurotransmitter molecules by fusion of the vesicle membranes and the cytoplasmic membrane. The transmitter diffuses rapidly across synaptic gap and stimulates an action potential at the sensitive surface of the postsynaptic fiber (if the concentration exceeds the threshold value).

neutrotransmitter. The transmitter rapidly diffuses (within 1 msec) across the synaptic gap and becomes attached to receptor molecules on the surface at the terminal end of the postsynaptic fiber; this affects the permeability of several ions and can result in a new action potential in the postsynaptic cell. When it has carried out its function, the transmitter is broken down enzymatically. Only very small threshold quantities (about 10^{-18} M or some 1000 molecules) are necessary to evoke an effect in the postsynaptic fiber.

Near the terminal end of the presynaptic fiber are small compartments containing neurotransmitter. These compartments are called *synaptic vesicles.* Upon the arrival of an action potential, the neurotransmitter is released from these vesicles by fusing of the vesicle membrane with cytoplasmic membrane of the neuron. Calcium ions seem to play an essential role in this process; the action potential opens up specific channels in the terminal and membrane, allowing Ca^{2+} to enter the neuron. In the cytoplasm of the neuron, Ca^{2+} then promotes the fusion of the vesical membranes and the cytoplasmic membrane.*

Not all synaptic connections between neurons evoke new action potentials. In many cases synapses result in an *inhibition* of the development of action potentials in the postsynaptic terminals. Such an inhibition is mediated by a special class of inhibitory neurons and involves specialized neurotransmitter molecules (see below). Inhibitory activity appears to be a very important factor of the functioning of neural networks that underlie behavior.

---

* This model was challenged recently; neurotransmitter, in the newly proposed model, is released from the neuron cytoplasm rather than from the vesicles (see Y. Dunant and M. Israël, The Release of Acetylcholine, Sci. Am., April 1985).

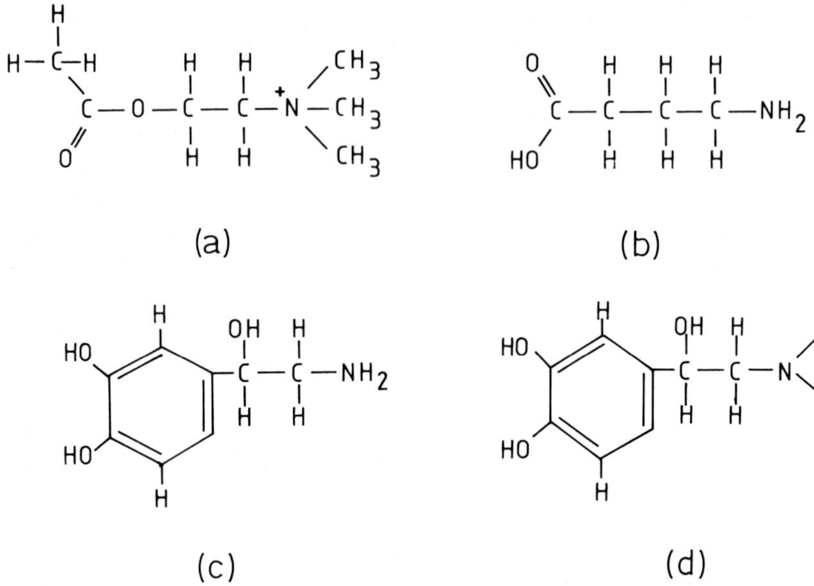

Fig. 9.7. The chemical structures of (a) acetylcholine, (b) gamma-aminobuteric acid (GABA), (c) norepinephrine, and (d) epinephrine.

Neurotransmitters, hormones, and neuropeptides

The longest known neurotransmitter is *acetylcholine* (Fig. 9.7a) a compound synthesized from the base choline, derived from phosphatidylcholine, and an acetyl group derived from acetylcoenzyme A. After its release into the synaptic cleft, it is broken down into acetate and choline by the enzyme acetylcholinesterase found at the outer surfaces of the cell membranes. Choline is recycled back into the presynaptic cell where acetylcholine is resynthesized and (at least partially) stored in the synaptic vesicles. The details of the action of acetylcholine are not known, but it is generally recognized that it is an excitatory transmitter in most cases.

An inhibitory transmitter is the amino acid *gamma-aminobuteric acid* (presently known by its acronym GABA (see Fig. 9.7b). It is not a component of proteins and peptides but it is synthesized, probably near the presynaptic terminal of specialized, so-called GABAergic neurons, by the enzyme glutamic acid decarboxylase. In mammals it is found mainly in brain tissue. GABAergic neurons have been best studied in the cortex of the cerebellum, an area of the brain which is responsible for the smooth coordination of muscular activity. The only neurons of the cerebellar cortex that have axons leaving the structure are the so-called *Purkinje cells*, so that other brain structures can be influenced only through the action of these cells. This action is inhibitory and the transmitter involved is GABA. Although it still is not quite understood how inhibitory action is involved in the control and the "fine-tuning" of motor activity, its importance is clear from this cerebellar arrangement. Another area in which inhibitory action and, for that matter GABA, plays a key role is vision. The directional sensitivity of visual inputs has a lot to do with the inhibitory action of specialized cells in the retina, the so-called amacrine cells (see Chapter 12).

Acetylcholine and GABA are not the only neurotransmitters. Many more are known presently, and their function seems not to be limited to communication between neurons. For example, *norepinephrine* and *epinephrine* (also known as noradrenaline and adrenaline, see Fig. 9.7c and d), which are derived from the amino acid tyrosine, are *hormones* as well as neurotransmitters. They are released from the adrenal glands and act as hormones to influence the heart rate and blood pressure and the flow of sugar in the bloodstream from the liver. They have a function as neurotransmitter in the autonomic nervous system. Another example are the *enkephalins* (there are two slightly different kinds), which, as hormones, regulate the movement of food in the intestines but are also neurotransmitters in the brain. The enkephalins are small peptides, five amino acids long; they belong to the larger group of *neuropeptides*. Most neuropeptides seem to have this dual function.

Both the synaptic system and the hormonal system are means of communication between cells. The main difference between the two systems is in the directness of action; the neurotransmitter acts very near the spot where it is released while hormones usually are released at locations away from their target cells. The fact that the functions of the agents of the two systems are interchangeable demonstrates, however, that the systems are closely related.

9.4 Information processing in neuronal systems

Action potentials thus are propagated and transmitted from neuron terminal to neuron terminal or to the endplates of muscle fibers. The transmission of information from neuron to neuron is based upon a binary code (the presence of absence of an impulse) but the code is not yet broken at all. The mechanism of the transduction of receptor signals to this binary code of impulses is also largely unknown (*cf.* Chapter 11). The transmission of the impulses occurs through complicated networks of neurons which are formed by literally millions of synaptic connections between axons and dendrites. This complication is aggravated by the fact that synapses can be both stimulatory and inhibitory. Moreover, electrophysiological studies have shown that different neurotransmitters can produce different effects at synapses. Also, a single neurotransmitter can have different effects depending on the type of synapse at which it is acting. An example of the latter is acetylcholine that closes potassium ion channels in the smooth muscle cells in the intestine (causing a gradual excitation of the muscle by an unknown mechanism) but opens sodium channels in skeletal muscle cells, causing a rapid and brisk contraction of the muscle. In the heart muscle, it inhibits the generation of an action potential.

In some cases, the release of a transmitter by the end of a presynaptic fiber may be below the threshold. In that case more subthreshold releases, hence the firings of more than one presynaptic fiber, are necessary to evoke a synaptic response. The possibility thus exists for an adding operation. Likewise, two, three, or more firings in a short time at one synapse may be necessary to produce a transmitted spike; then the synapse acts as a divider. Multiplication can be achieved when several terminals from one neuron have synaptic connections, with different time delays, at the same second neuron. Subtraction occurs when the released transmitter in a synapse acts as an inhibitor rather than a stimulator. Such synapses have been found in the heart muscle; acetylcholine, which produces an action potential at motor endplates and other synapses inhibits the generation of a spike in heart muscle. At such synapses the transmitter increases the permeability to K^+ and larger cations but does not change the Na^+ permeability at all. The net result is a change in the membrane potential and an increase in the firings at other synapses necessary to transmit the spike. This effectively results in a subtraction of impulses from two different incoming neurons.

The nervous system thus operates with a circuitry not unlike that of a modern digital computer, only far more complex and highly nonlinear. In addition, the synaptic conduction is influenced by slow potential fluctuations. Not only the K^+ and Na^+ concentrations but also the concentrations of Ca^{2+} and Mg^{2+}, and in particular their ratio, can alter synaptic conduction. It has been shown that at the neuromuscular synapse the probability of a given packet of about 1000 molecules of acetylcholine (the threshold quantity) entering the intercellular fluid is a function of the Ca^{2+}/Mg^{2+} concentration ratio.

Many, in fact most, aspects of the functioning of nervous systems, nerves and nerve conduction are still not understood. Although we have a notion now of how an action potential can arise, we still have no idea how transmitter molecules, or other membrane disturbing agencies, can change the ion permeabilities. Still larger looms the problem of understanding the way in which the information from the environment is coded and processed. More knowledge about this may eventually lead us to the understanding of things like perception, memory, and learning.

Bibliography

Bezanilla, F., Vergara, J., and Taylor, R.E. (1982). Voltage Clamping of Excitable Membranes, in: "Biophysics" (G. Ehrenstein, and H. Lecar, Eds.), "Methods of Experimental Physics, Vol. 20", pp. 445-511, Acad. Press, New York.

Bloom, F.E. (1981). Neuropeptides, Sci. Am.245 (October), 114–124.

Grillner, S. (1985). Neurobiological Basis of Rhythmic Motor Acts in Vertebrates, Science 228, 143–149.

Hökfelt, T., Johansson, O., Ljungdahl, Å, Lundbergh, J.M., and Schultzberg, M. (1980). Peptidergic Neurons, Nature 284, 515–521.

Poggio, T., and Koch, C. (1987). Synapses that Compute Motion, Sci. Am. 256 (May), 42–48.

Snyder, S.H. 1985). The Molecular Basis of Communication between Cells, Sci. Am. 253 (October), 114–123.

Thompson, R.F. (1986). The Neurobiology of Learning and Memory, Science 233, 941–947.

Wurtman, R.J. (1986), Nutrients that Modify Brain Function, Sci. Am. 246 (April), 42–51.

Zucker, R.S., and Lando, C. (1986). Mechanism of Transmitter Release: Voltage Hypothesis and Calcium Hypothesis, Science 231, 574–579.

10. Biophysics of contractility

10.1 Intracellular motion

Mechanical work

Mechanical work is performed by all living organisms. Moving of cells or cell aggregates (organs like muscles or organisms as a whole) is the most obvious form of mechanical work; it gives organisms the ability to move away from a noxious environment or to move toward beneficient regions of space. However, movements *inside* the cells are also forms of mechanical work performed by most if not all living cells. These include processes like transport of particles and organelles in different directions within the cell, protoplasmic streaming, mitosis (cell division), swelling and contraction of organelles, and endocytosis. Common factors prevail in all these cases; first, the energy source is always the free energy of the hydrolysis of ATP to ADP and inorganic phosphate. Furthermore, most mechanisms share a common underlying molecular basis. Wherever cell movement is studied on the molecular level, it turns out to involve one particular class of proteins of which actin and myosin are the best known. These proteins, or combinations of them, generally exhibit ATPase activity, which clearly indicates their importance in the transduction of chemical energy into mechanical work.

Cytoskeletal motion

Within a cell there is a continuous motion of particles, vesicles and even organelles such as mitochondria. The first clear historical account of these processes is that of the 19th-century naturalist and microscopist Joseph Leidy, who observed such movements in the tentacle-like protrusions of a unicellular organism called *Gromia*. He described the movements as a dynamic stream of motion that seems to consist of ".. pale granual protoplasm with coarser and more defined granules .....in incessant motion along the course of threads..". The application of improved techniques, such as video enhanced contrast microscopy, recently discovered by R.D. Allen, showed that the motion indeed seems to be "guided" by a network of linear elements in the cell and that these elements appeared to be the microtubules, which are part of the cytoskeleton (see Chapter 2).

Microtubules, the 25 nm thick filaments that contain the protein *tubulin* and that extent throughout the cell, are essential for this process. Each of these tenuous threads can move particles in two directions simultaneously. They also move themselves, usually in only one direction until they meet an obstacle after which they bounce off at a certain angle or "fishtail" in a regular series of curled patterns.

The functioning of this transport system can be seen clearly in the axons of neurons. Images obtained by video-enhanced contrast microscopy of the giant axon, which is part of a large transparent neuron in the squid, showed that vesicles that appeared to be precursors of the synaptic vesicles, move from the cell body in the direction of the presynaptic terminal; neurotransmitter that is synthesized by the Golgi apparatus thus is transported for release at the axon terminal upon arrival of an action potential. Other multivesicular bodies containing membrane debris are moved in the opposite direction, towards the cell body, where lysosomes take care of degradation and disposal. Mitochondria move in both directions, supplying energy where it is needed. Transport of the vesicles appeared to be continuous; that of the mitochondria is intermittent.

It seems to be clear that a single microtubule can support transport in both directions. The mechanism by which these movements occur, however is largely unknown. ATP is an essential requirement. There is some microscopic evidence that the movements are caused by conformational changes of cross-bridges, or so-called sidearms, which have sites with ATPase activity. The sidearms bind ATP, thus becoming activated. ATP hydrolysis then changes the conformation of the sidearms in such a way that some sort of a crawling action results. Such a mechanism has similarities with the mechanism con-trolling contraction of skeletal muscles (see section 10.3). It is not clear, however, whether the sidearms, before activation, reside on the microtubules, on vesicles and organelles, or are present in free form. In activated form they must contact the microtubules, because no movement independent of the microtubules has ever been detected.

10.2 Cellular motion

Cilia and flagella

Cells that are able to move through a liquid medium or to move the medium across their surface do so by means of whiplike structures called *cilia* or *flagella*. Cilia are short fibers occurring in bundles. Flagella are longer and usually occur in much smaller numbers together or as single fibers. Eukaryotic cilia and flagella are all built according to the same regular patterns and are of about the same diameter (about 0.5 μm). Figure 10.1 shows the flagellum of a sperm cell. The flagellum originates in a basal granule, or midpiece, which is found in the cytoplasm of the cell. The basal granule is covered with long mitochondria which are helically wound around the granule. The flagellum itself consists of nine pairs of outer fibrils around two inner tubules. Extractions of flagella show the presence of a complex protein very similar to the complex actomyosin found in muscle (see section 10.3). It seems probable that the outer fibrils are composed of actomyosin; the inner tubules could serve as channels

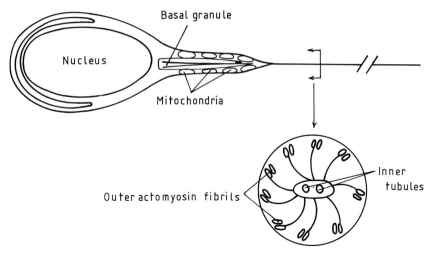

Fig. 10.1. A sperm cell with flagellum. The flagellum originates from the basal granule and shows the nine double actomyosin fibers and the set of two inner tubules in cross section.

for the diffusion of ATP to be hydrolyzed at the actomyosin sites.

The cells move by sweeping beats of the flagella or cilia, sometimes with a slightly circular component like the cilia of protozoa, or in a single plane, like those of the epithelium cells of the human respiratory tract. Flagella of sperm cells beat in a wavelike fashion. The regular arrangements of the fibers in the cilia and flagella suggest that ciliar or flagellar beat could be based on the contraction of the fibers on one side of the array, drawing the tip of the cilium in that direction. Thus, the ciliar or flagellar movement to be accounted for is rather complicated. No theory, so far, has been put forward which would give a satisfactory explanation for the way these movements are produced.

Bacterial flagella do not show the tubular array of the eukaryotic flagella or cilia. They are much smaller, about 0.15 μm in diameter, and in electron micrographs show a helical structure with a pitch that seems typical of each particular species. The mechanism of bacterial flagellar motion is not known, although their movement is less complicated than that of eukaryotic cells. The photosynthetic bacteria *Rhodospirillum rubrum*, for example, has flagella at each polar end. When the cell moves the flagella rotate around the cell describing a cone, so that the bacterium rotates in the opposite direction. Since the bacterium is a spiral itself, its rotation drives it through the liquid medium. The cell can reverse direction very rapidly by simply flipping its flagella at each polar end in the appropriate direction in apparent coordination with each other. This reversal of direction occurs during *phototaxis*, the light-induced movement of many photosynthetic organisms (see Fig. 10.2). Crossing boundaries from light to darkness causes *Rhodospirillum rubrum* cells to abruptly change their swimming direction. If a cell crosses the boundary from darkness to light no reaction occurs. The cells, thus, have the tendency

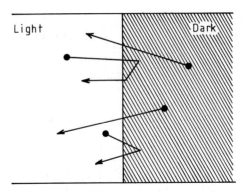

Fig. 10.2. Phototaxis of the photosynthetic bacterium *Rhodospirillum rubrum*. The cells reverse direction when crossing the boundary from light to dark. No change in direction occurs when the cells cross the boundary from dark to light.

to accumulate in the light. The light-induced synthesis of ATP in photophosphorylation is connected with this motile reaction.

10.3 Muscular contraction

Muscles

In higher organisms of the animal kingdom, the capacity to perform mechanical work is exhibited by highly specialized structures called *muscles*, groups of cells which move as a unit by means of contracting intracellular filaments. Histologically, we can distinguish two types of muscles: the *smooth muscles*, tissue which forms the walls of internal organs like intestines, blood vessels (particularly the arteries), esophagus, and bladder, and the *striated muscles*.

Smooth muscle consists of spindle-shaped cells about 0.5 mm long and about 0.02 mm diameter (Fig. 10.3). Contraction of smooth muscle is not under voluntary control and proceeds in a slow and generalized manner. The *cardiac muscle* (the myocardium or heart muscle) and the *skeletal muscles*

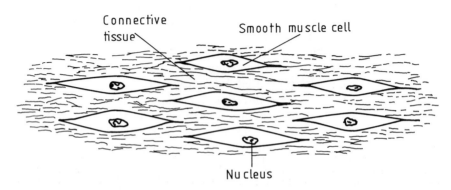

Fig. 10.3. Smooth muscle cells.

242

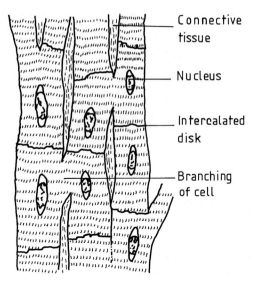

Fig. 10.4. Cardiac muscle.

are striated. The cells of cardiac muscle are fibrous (Figure 10.4); its periodic contraction and relaxation goes on automatically, although the sympathetic nervous system can influence the rate of the heart beat. Most of what is known about cellular motion comes from studies with skeletal muscle and it is the only contractile organ about which a notion of the mechanism of contraction exists. Striated muscle, cardiac muscle as well as skeletal muscle, is highly

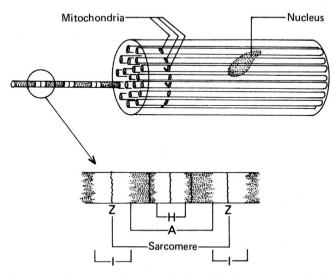

Fig. 10.5. The composition of a striated muscle fiber. The fiber consists of a bundle of many myofibrils which have the striped appearance. The myofibril is a lineup of many sarcomeres; in the sarcomere one can distinguish the isotropic I bands, the anisotropic A bands, the Z lines, and the central H disk.

243

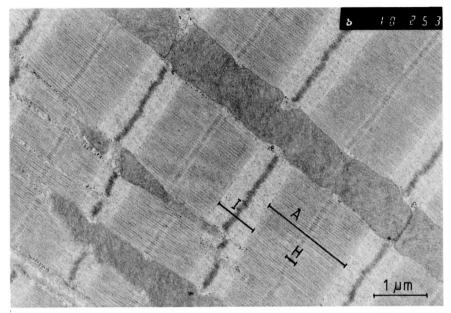

Fig. 10.6. Electron micrograph of a section through the myocardium of a dog. The sarcomere structure is clearly visible. The I bands, optically isotropic regions, are at each side of a Z line. The darker regions at each side of an I band are the optically anisotropic A bands. In the middle of an A band is the H disk with a sharper M line. (Courtesy of Dr. W. Jacob and Dr. F. Lakiere, Electron Microscopy Laboratory, Universitaire Instellingen Antwerpen, Antwerp, Belgium).

organized and has many elongated cells, the muscle fibers, which contain the contractile organelles, the myofibrils (see Fig. 10.5). Myofibrils show characteristic repeating structures, about 2.5 μm in length, which give them the striated appearance. The repeating unit is called a *sarcomere* and is clearly visible in the electron micrograph of part of the myocardium of a dog, shown in Fig. 10.6.

The sarcomere
A sarcomere has a banded structure formed by regions of sharply different optical properties. The boundaries of the sarcomere, the *Z lines*, run in the middle of light bands which are called *I bands* because they are optically isotropic regions. At each side of an I band is a darker region which shows strong birefringence. This is the *A band* (A for anisotropic). The middle of the A band is somewhat lighter, forming an *H dis* within which a sharp darker line, the *M line* can be seen.

This optical appearance is caused by an array of filaments which make up the different bands. Figure 10.7 shows a schematic drawing of a blown-up sarcomere. There are two kinds of filaments; the *thick filaments* making up the A band, and the *thin filaments* responsible for the I band which are attached at one terminal end to a plate called the Z line. Where the thick and thin filaments overlap is found the darker part of the A band. The H

244

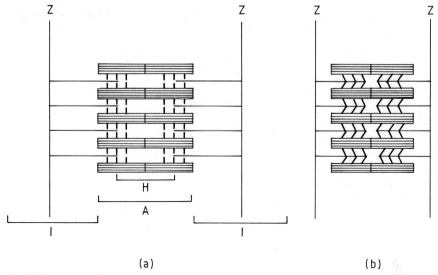

Fig. 10.7. A schematic representation of a sarcomere; (*a*) muscle relaxed and (*b*) muscle contracted. The bars represent the thick filaments, the lines the thin filaments connected to the thick filaments by cross-bridges. The H disk is the area with only thick filaments, the A band includes the H disk and the overlap area between the thick and the thin filaments. The I band is the area in which there are only thin filaments. The sliding-filament model considers contraction as resulting from a sliding of the thick and thin filaments along each other, thus shortening the distance between two Z lines and causing the I bands to disappear.

disk is the region where only thick filaments are found. The array is extremely regular in space. Each thick filament is surrounded, in the region of overlap, by six thin filaments (see Fig. 10.8). The filaments are connected to each other by tiny, so-called cross-bridges.

The sliding-filament model
Carefully designed experiments, in which observations were made with an electron microscope, and optical measurements have shown that when the muscle contracts the I zones shrink and finally disappear in the fully contracted muscle; the H disks also disappear. These observations have led to the *sliding-*

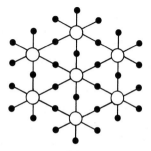

Fig. 10.8. The extremely regular arrangement of the thick (open circles) and the thin (dots) filaments in a cross section of a sarcomere.

filament model for the contraction mechanism. This model, which does have substantial molecular justification considers contraction as resulting from a sliding of the filaments along each other, thus making the region of overlap of the two kinds of filaments larger. The sliding of the filaments could be visualized as being caused by a repeated detachment and reattachment of the cross-bridges between the two kinds of filaments. The cross-bridges change angle during each cycle of attachment and detachment, pulling a thick filament along a thin one.

The sliding-filament model is consistent with all information available thus far about muscular contraction and muscular morphology. For instance, when a muscle is stretched it loses the ability to exert maximum force. This result is a direct consequence of the geometry described above. The force of the muscle, according to the sliding-filament model, depends on the interaction of the two kinds of filaments and this interaction depends on the degree of overlap. In stretched muscle, the overlap is diminished and thus the force is lessened.

Contractile proteins
The two major protein components in myofibrils were found to be *myosin*, which makes up about 50%-55% of the total protein, and *actin*, which accounts for 20%-25%. Actin is the major protein of the thin filaments. When it is extracted from the muscle it comes out as a globular protein, the so-called G-actin, with a molecular weight of about 60,000. When G-actin is incubated with ATP in the presence of Mg^{2+} the protein polymerizes to a larger aggregate, the fibrous F-actin. In the muscle fibril the actin globules are arranged in double helical strands like two intertwined strands of beads (Fig. 10.9). Two other proteins are closely associated with actin in the thin filaments, *tropomyocin* and *troponin*. Tropomyosin is a long thread-like molecule. The tropomyosin molecules attach end to end to each other in long filaments on the surface of an actin strand. Each actin strand carries its own filament which lies near the groove between the two pairs of strands. Each tropomyosin molecule covers seven G-actin monomers. Troponin, on the other hand, has a more globular shape. It has been shown that troponin consists of three subunits. It is attached to the tropomyosin filament, one on each molecule of tropomyosin, at about a third of the way from the end. The tropomyosin-troponin system plays an important role in the regulation of contraction by Ca^{2+} ions.

The thick filaments contain all the myosin of the muscle. Myosin is a large molecule with a molecular weight of about 500,000. This protein appears

Fig. 10.9. A thin filament consisting of two intertwisted strings of globular actin molecules. The troponin molecules (black ovals) and tropomyosin molecules (lines near the grooves between the two actin strands) are seen around the actin filament. Each tropomyosin molecule is connected to one troponin molecule, covering seven globular actin molecules.

246

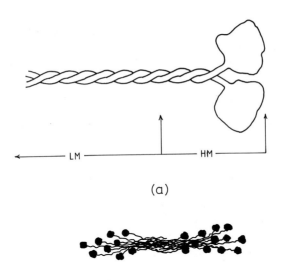

(a)

(b)

Fig. 10.10. A schematic representation of a. a myosin molecule; and b. a bundle of myosin molecules as found in the thick filaments. The myosin molecule consists of two intertwined helices each with a head piece. The arrow represent the site at which the molecule is cleaved into light meromyosin and heavy meromyosin by tripsin.

to be an elongated filament, consisting of two intertwined helices, each with a globular head (Fig. 10.10a). A thick filament of a normal muscle contains several hundred myosin molecules (Fig. 10.10b). The molecule can be cleaved at a specific place by a protolytic enzyme (trypsin), yielding a light fraction called *light meromyosin*, which is the filamentous end of the molecule, and a heavy fraction called *heavy meromyosin*, which contains the globular heads. The heads can be isolated from the tail by the treatment of myosin with the enzyme papain. Myosin has ATPase activity and the cleavage experiments have shown that this ability to hydrolyze ATP is located at the globular heads. The heads also appear to be the site of greatest affinity for actin when the two proteins mix and form a complex. This strongly indicates that the globular heads are the part of the myosin molecule that form the cross-bridges with actin in the myofibril.

Actomyosin
In vitro, the two proteins form a complex when they are mixed together: this complex is called *actomyosin*. Complexes are also formed when heavy meromyosin or isolated globular heads are used instead of myosin. The complexes can be used as model systems in order to find out more about the properties of the system. The complex, for instance, can be precipitated into an insoluble sheet that contracts upon addition of ATP. The ATP is actually hydrolyzed in the experiment, clearly showing the coupling of ATPase activity to the performance of mechanical work.

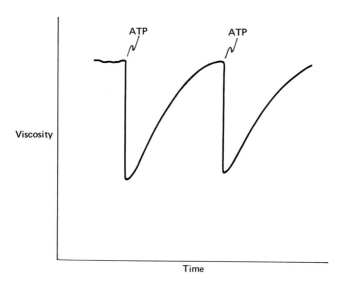

Fig. 10.11. An experiment showing the drastic change of the viscosity of a suspension of actomyosin upon addition of ATP. The binding of ATP to the actomyosin complex causes dissociation into actin and myosin-ATP complex. When ATP is hydrolyzed actin and myosin recombine into actomyosin and the viscosity increases.

Contraction mechanism

Most suggestive for the sliding-filament model is that the binding of ATP to actomyosin leads to a sharp decrease in the viscosity of a suspension of the complex. The effect, however, is reversible (as shown in Fig. 10.11). Addition of ATP to a suspension of actomyosin which is quite viscous and shows birefringence, causes a sharp decrease of the viscosity. When the ATP is being hydrolyzed (in the presence of Mg^{2+}) a slow recovery to the original viscosity occurs, after which the experiment can be repeated. It has clearly been shown that the viscosity change in the presence of bound ATP is the result of a dissociation of the actomyosin into free actin and myosin·ATP complex. It thus follows that not the *formation* but the *breakage* of the cross-bridges between the myosin heads and the actin in the thin filaments requires the binding of ATP.

In the muscle, the interaction between the two kinds of filaments during contraction is a cyclic sequence of steps which can be summarized as follows (Fig. 10.12):

1. Binding of ATP to the myosin heads.
2. Activation and binding of the activated myosin·ATP complex to actin in the thin filaments; this reaction occurs only in the presence of CA^{2+} ions.
3. Hydrolysis of ATP.
4. Release of phosphate, which changes the angle of the myosin heads in respect to the myosin filament from 90° to 45°.
5. Release of ADP and binding of ATP to the myosin heads, dissociating

248

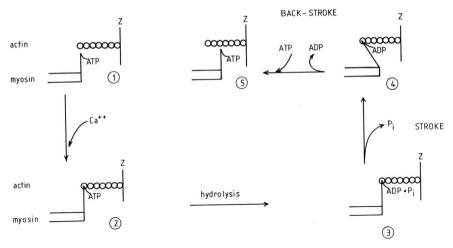

Fig. 10.12. A schematic representation of the mechanism of contraction: ATP binds to the myosin heads and the myosin·ATP complex is activated (1); Ca^{2+} then triggers the binding of the myosin·ATP complex to actin (2); ATP hydrolyzes (3); the release of phosphate results in a change of angle of the myosin heads in respect to the thick filaments from about 90° to about 45° (4); this pulls the thick filament along the thin filament, in a stroke, not unlike that of an oar pulling a rowboat through the water; ADP is released and the binding of another ATP molecule to the myosin heads then causes the dissociation of the actomyosin complex (5), like the backstroke of an oar; the cycle then is repeated from step (2).

the myosin·ATP complex from the actin filament, after which the cycle is repeated from step 2.

This sequence of steps is in accordance with the biochemical and biophysical evidence available at the time of this writing. It should be emphasized that not the hydrolysis itself of ATP but the release of the phosphate is associated with the most important change in free energy. During that step force is applied and mechanical work is done. This can be understood by realizing that the free energy of ADP + phosphate in solution is much lower than the free energy of ATP in solution. It is this mode of action of ATP that also can explain the energy-linked change in conformation of transport enzymes like the Na^+K^+ATPase and the reversible ATPase of the coupling factor (see Chapters 7 and 8).

One should realize that the mechanical work that is performed by a contracting muscle is not that of single myosin heads separately but that of a great number of heads together. The transition, in step 4, from the 90° conformation to the 45° conformation of the heads is constrained by the filament lattice. The result of step 4, therefore, is not a stable 45° cross-bridge but a *strained* 45° cross-bridge that exerts a (positive) force. The strain is gradually relieved, and work is performed, only as the filaments slide past each other. The same applies, in the opposite sense, to the "back-stroke". Also the 90° cross-bridge is strained, applying a *negative* force. This does not result in a reversal of the original stroke because the detachment of the

249

myosin·ATP complex from the actin is too rapid for the force to do negative work. The latter is a basic assumption underlying the model.

According to the model just described, the actomyosin complex can exist in two forms, one activated and the other inactivated. The activated form is generated when Ca^{2+} triggers the binding of activated myosin·ATP to the thin filaments. The inactivated form is that part left after the hydrolysis of the ATP, which, in the absence of ATP, is very stable. This is easily produced in the laboratory by mixing myosin with actin in the absence of ATP. Binding of ATP. Binding of ATP, however, immediately causes the detachment of the myosin·ATP complex from the thin filaments, as shown by the viscosity experiment described above. The stability of the inactivated actomyosin complex accounts for the well-known phenomenon of *rigor mortis*, the extreme rigidity that develops in the muscles after death. The gradual disappearance of ATP following death more and more prevents the binding of ATP to the myosin heads, thus leading to an increasing amount of inactivated complex. The inactivated complex, therefore, is sometimes called the *rigor complex*.

Role of Ca^{2+}

There is evidence that strongly suggests that the onset of contraction is mediated by Ca^{2+} ions and that this Ca^{2+} trigger acts through the tropomyosin-troponin system. Contraction and hydrolysis of ATP in suspensions of actomyosin prepared from muscle is dependent on the presence of Ca^{2+}. If, instead of the thin filaments, a purified preparation of F-actin is used, contraction becomes insensitive to Ca^{2+} and ATP hydrolysis goes on randomly until all of the ATP is used. From these and other results it appears that the two proteins tropomyosin and troponin are required for the control of the contraction mechanism and that Ca^{2+} is the agent by which such control is exerted.

Of all the components of the contractile system in striated muscle only troponin can bind calcium. All available evidence shows that the step that is sensitive to Ca^{2+} is the binding of the activated myosin·ATP complex to actin (step 2). It looks then as if the two proteins tropomyosin and troponin are responsible for blocking a site on the actin for the binding of the activated myosin·ATP and that the binding of Ca^{2+} to troponin causes the site to be exposed so that binding of the activated myosin heads can occur. A model in which these assumptions are incorporated is shown in Figure 10.13. In the absence of calcium the troponin molecules hold the tropomyosin strands on the actin filament in such a way that a site for the binding of the activated myosin heads is blocked. The binding of Ca^{2+} to the troponin causes conformational changes in the molecule in such a fashion that the tropomyosin thread is pushed deeper into the grooves of the actin strands, away from the blocking site (thus allowing the activated myosin heads to "catch in"). The binding of the Ca^{2+} ions can be pictured as leading to a "tightening" of the bonds among the subunits of the troponin and a weakening of the interaction of the molecule with actin; lowering the calcium level "loosens" the troponin complex and makes it bind more strongly to actin. The tropomyosin strand amplifies this action over seven actin molecules. X-ray

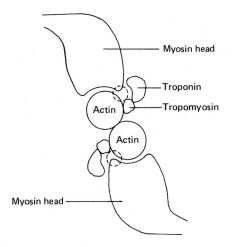

Fig. 10.13. A possible model for the Ca^{2+} trigger of muscle contraction. The binding of Ca^{2+} to troponin causes a conformational change in the molecule which results in the tropomyosin being pushed deeper into the grooves between the two actin strands, thereby exposing the binding site for the binding of the myosin head to the actin.

diffraction studies of troponin and tropomyosin under different binding conditions have supported this model.

10.4 The energetics of contraction

The energy source of contraction

It is presently beyond all doubt that ATP is the direct source of energy for the contractile process. This is also the case *in vivo*, although it does not appear so at first sight. Muscles can also contract under complete anaerobic conditions or after poisoning with cyanide blocking respiratory electron transport. ATP then can be supplied by glycolysis and, indeed, the glucose concentration decreases with a proportional increase of lactic acid in these cells. But even when glycolysis is inhibited, for instance by inhibiting the enzyme phosphoglyceraldehyde dehydrogenase by iodoacetate, contraction of the muscle still goes on. Subsequent chemical analysis of the muscle reveals that the amount of ATP in the cell has remained pretty much constant but that the concentration of another "energy rich" phosphate, phosphocreatine, has declined considerably with a proportional increase of the concentration of creatine. Phosphocreatine by itself, however, cannot make muscles contract if there is no ATP or ADP present. Thus, it looks as if phosphocreatine is providing a supply of ATP and, indeed, the only known enzymatic reaction in which this compound participates is

$$\text{phosphocreatine} + \text{ADP} \rightleftarrows \text{creatine} + \text{ATP}. \tag{10.1}$$

a reversible reaction catalyzed by the enzyme creatine phosphokinase. We have already seen in Chapter 5 (Table 5.1) that the standard free energy change of hydrolysis of phosphocreatine is substantially more negative than that of ATP. Equilibrium of reaction 10.1 is, therefore, shifted to the right. Moreover, the use of ATP in muscle contraction tends to diminish the concentration of ATP, thus letting reaction 10.1 proceed to the right and keeping a constant level of ATP. Poisoning the creatine phosphokinase by dinitrofluorobenzene results in a rapid decline of the ATP concentration and a constant level of phosphocreatine.

Phosphocreatine is regenerated from creatine and ATP; the latter is formed by oxidative phosphorylation during the recovery period. The phosphocreatine pool, thus provides for a reservoir of "high-energy" phosphate groups; the ATP concentration in the muscle is always maintained at a high and constant steady-state level by this system. The system allows the muscle to operate even in a short time when not enough oxygen is present to produce sufficient ATP.

All these arguments provide plenty of evidence for the direct involvement of ATP in muscle contraction, *in vivo* as well as *in vitro*. This is, of course, entirely consistent with the universal role of ATP in biological energy transformations and, for the muscle in particular, with the striking juxtaposition of numerous mitochondria to the myofibrils.

Ca^{2+} and the muscle synapse

Skeletal muscle contracts upon stimulation by a motor nerve. The communication between nerve and muscle cell, as we have seen in the previous Chapter, is a synapse between the nerve end and an endplate or *myoneural junction*. The endplate is a postsynaptic surface, sensitive to the synaptic transmitter acetylcholine. Smooth muscle as well as heart muscle are contracting in an autonomous way; they contract rythmically even when all nerve connections are removed.

It seems clear that in skeletal muscle the initiation of the contraction is mediated by Ca^{2+}. How Ca^{2+} becomes available for the control of contraction is in fact unknown, although some suggestive hypotheses do exist. The stimulation of the motor nerve causes the release of acetylcholine in the synapse between nerve and endplate. This produces a local potential change at the endplate surface which, when the threshold value is exceeded, causes a rapidly spreading action potential in the cell membrane of the muscle cell, and the cell contracts. It seems that changes in the ion distributions across the cell membrane, in particular that of Ca^{2+}, caused by the action potential make the connection between the action potential, which is a property of the cytoplasmic membrane, and the contractile process, which is a property of the myofibrils. It is likely that the action potential at the cytoplasmic membrane is linked to the release of Ca^{2+} in the immediate region of the contractile proteins, by means of a system of tubules which is continuous with the plasma membrane as well as the endoplasmic reticulum. This tubule system is called the *sarcoplasmic reticulum* and the close proximity of this system of channels

252

to the contractile apparatus leads to the possibility that the release of Ca^{2+} is "felt" simultaneously by the entire bundle of myofibrils, so that they can contract synchronously.

Bibliography

Allen, R.D. (1987). *Op. cit.* (Chapter 2).
Cohen, C. (1975). The Protein Switch of Muscle Contraction, *Sci. Am.* **233** (November), 36–45.
Eisenberg, E., and Hill, T.L. (1985). Muscle Contraction and Free Energy Transduction in Biological Systems, *Science* **227**, 999–1006.
Weber, K., and Osborn, M. (1985). *Op. cit.* (Chapter 2).

11. Biophysics of the sensory systems

11.1 The transmission of information

The senses

Multicellular organisms, especially the higher animals, are stimulated by the environment through sensory systems; specialized organ systems which transmit information from the environment to the brain. We traditionally speak of them as the five senses: vision, hearing, olfaction, taste, and touch. But they, in fact, include more. Sensations of pain and temperature, and even sensations of hunger and thirst can be thought of as being evoked by sensory detectors. We can, in general, classify sensory detectors as follows. First, there are the special senses of vision and hearing involving the highly specialized organ systems of the eye and ear. These are called *teleceptors*, receiving information from distant objects. Second, we have the *chemical receptors*, those receptors that are excited by chemical stimulation; olfaction and taste belong to this group. Third, *somatic receptors* are those responsible for the sensations of the body, namely touch, pressure, pain, and temperature. Finally, there are the *visceral receptors* that keep the brain informed about conditions inside the body and are responsible for the sensations of hunger, thirst, and the urge to discharge.

By means of the senses, organisms are able to respond appropriately to outside stimuli: there is a coordination between sensory information and muscle action (we jump out of the way if we see an oncoming automobile; we turn away if something smells bad; we cover our ears if we hear loud noises). Dramatic examples of more refined coordination are someone doing a fine piece of woodcarving or someone driving an automobile. Apparently, the senses are part of complicated *feedback control systems*.

Neuronal coding of receptor signals

Although the senses are quite different from each other, as are the results of the different types of sensory perception (we know very well the difference between what we see and what we hear or smell), there are some common features which justify a generalized discussion. An important aspect of the senses is that they are all *transducers*: they transform one form of energy into another, or rather, translate information from one code into another. Information contained in electromagnetic radiation (vision), mechanical vibration (hearing), mechanical pressure and temperature (touch), and chemical structures (olfaction and taste) is translasted into a code that can be handled

255

by the nervous system. This information is collected in *receptors* or *receptor organs*, where the translation is then carried out. The receptor organs for vision and hearing, the eye and the ear, are highly specialized and quite elaborate. The interaction of electromagnetic radiation and the receptor cells in the eye is reasonably well-known (see section 11.2). The state of our knowledge of the interaction of sound waves and the auditory parts of the ear (see section 11.3) is less certain. In both cases, however, the transduction process proper is still an unsolved problem. The receptors for olfaction (the olfactory lobe in the upper part of the nose), taste (the taste buds in the tongue), and the receptors for touch, pain, and temperature are less elaborate. The transduction process in these cases is also unknown.

When stimulated, each of these receptors will initiate nerve discharges in the form of action potentials. Each of them, of course, is particularly sensitive to specific stimuli, such as light for the eye and sound for the ear. Nerve discharge, however, is also initiated in each receptor when it is subjected to excessive pressure, damage, or high temperature. The resulting sensation, however, is always that characteristic for the particular receptor, even when the stimulus is abnormal. The common experience of "seeing stars" when the eyes are struck is an example of this. Thus, it seems that the recognition of a particular kind of stimulus by the brain is linked to the origin of the stimulus, rather than to a particular kind of code transmitted through the nervous system. The answer to the question of how the brain does distinguish stimuli from one type of receptor from that from another type of receptor is not easily discerned. The anatomical relation between certain loci in the brain and the sensory organs is only part of the answer. The high degree of processing that is going on in the organs themselves as well as in the brain contributes to the tremendous complexity of the problem.

The code transmitted through the nervous system is a binary code; as stated in Chapter 9, action potentials are "on-or-off" events. This fact would make concepts developed in disciplines like information theory and cybernetics applicable to this subject. After all, digital computers, the rapid development of which would have been unthinkable without information theory and cybernetics, operate with a binary code, make use of electrical pulses, and are systems of networks in which feedback control is essential. Indeed, concepts of information theory and cybernetics have been used, but so far with only limited success. It has to be borne in mind that both of these disciplines do not treat *things* but *ways of behaving*; they do not ask "what is this thing" but "what does it do". They are sciences in their own right, not depending on or being derived from any other branch of science such as physics or biology. Therefore, they never can provide answers to questions about mechanisms. One could still hold, however, that applications of cybernetics, and perhaps even more so with information theory, help express ideas and hypotheses and, more important, help to realize similarities and analogies between widely diverse fields. In such a way it can lead to a direction in which certain explanations must go. In itself, however, it never can offer such explanations or lead to new hypotheses about mechanisms.

256

Sensory signal processing

As we have stated before, the way in which sensory information, such as the image on a retina or a tone of a certain pitch at the entrance of the cochlea, is translated into the binary code of action potentials is unknown. For higher, more developed organisms it is certain that, with perhaps the exception of the sense of touch, it is not by a simple correspondence between a receptor cell and a neuron. In the case of hearing, for example, the old Helmholz theory in which tuned resonators in the cochlea were thought to excite their own neuron cannot hold because the frequency range of hearing (spanning 10 octaves from 20 Hz to 20 kHz) is orders of magnitude larger than the maximum firing frequency of a neuron (less than 1000/sec.). It seems clear that the cooperation of many neurons is necessary in order to cope with the frequency capacity of the auditory sense.

This is even more obvious with vision, in which the cooperation of many nerve cells has actually been demonstrated. Figure 11.1 shows a schematic drawing of the arrangement of cells in the retina. Receptor cells (rods and cones) form synaptic connections with intermediary cells which in turn have synaptic connections with ganglion cells, the axons of which together form the optic nerve (for a complete picture of the eye, see Fig. 11.4). The intermediary cells can be distinguished as *horizontal cells, bipolar cells*, and *amacrine cells*. These cells form a network of communication, processing the visual information and delivering it to the optic nerve. The signals from the receptor cells are transmitted to the bipolar cells which in turn stimulate the ganglion cells, either directly or through the amacrine cells. The horizontal cells make cross connections among receptor and bipolar cells, and the amacrine cells provide contacts between bipolar cells and ganglion cells. The synapses

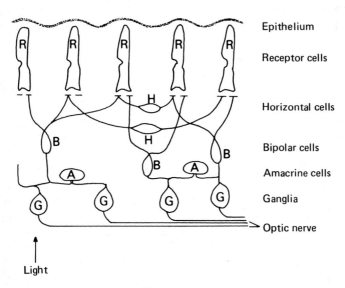

Fig. 11.1. A schematic drawing of the different layers of cells and their interconnections in a vertebrate retina.

257

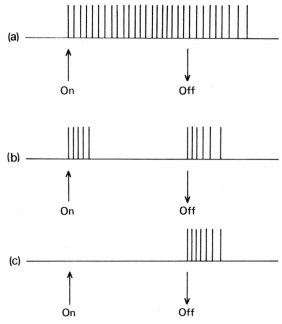

Fig. 11.2. Three patterns of responses of different fibers in a single optic nerve. (a) the "on-response" is a volley of action potentials triggered by turning on the light without any effect by turning the light off. (b) The "on-off-response" is a short burst of action potentials when the light is turned on and a similar burst when the light is turned off. (c) The "off-response" is a volley of action potentials only when the light is turned off. These responses are a result of a combination of excitatory and inhibitory synaptic connections.

can be stimulatory or inhibitory. In this way, one ganglion cell is "served" by a number of receptor cells. The area covered by these receptor cells is called the *receptive field* of the ganglion.

One can map the receptive field of a ganglion by measuring the action potentials evoked in a single fiber of the optic nerve by moving a tiny spot of light across the surface of the eye. A number of experiments of this type have been carried out with vertebrates as well as invertebrates. The results have shown that, in general, three patterns of recordings of action potentials can be found in different fibers of a single optic nerve (Fig. 11.2). These on, on-off, and off responses are the basis of a complicated information processing system and the basis of many different patterns of response. Movement and contrast detection involve different ganglions. Many ganglions, movement detectors as well as contrast detectors, show a pattern of antagonism between central and peripheral regions of the receptive field. When illumination of the center of the receptive fields evokes an "on" response in such a ganglion, additional illumination of the peripheral region suppresses it. The reverse can also be true. It seems that the inhibitory action against either an "on" response or an "off" response involves messages reaching the connecting bipolar cell from a receptor cell through intervening horizontal cells.

Most of the details of the signal manipulation of this network of nerve cells remain obscure. It is clear, however, that transmission and processing of sensory information, at least as far as vision and hearing are concerned, occur at the level of the receptor organs as well as in the brain itself.

Generator potentials

The action potentials generated in the sensory nerves (the optic nerve, the auditory nerve, the olfactory nerve, the dendrites inside the taste buds, and perhaps also the fiber endings connected to the receptors for pressure, temperature, and pain) are usually evoked by synaptic stimulation. A presynaptic cell may itself carry action potentials or it may only generate a change in its membrane potential, lacking the regenerative and traveling aspects of an action potential. These slower "generator potentials" are often observed in cells comprising sensory systems. They have magnitudes that vary with the intensity of the stimulation of the receptors, although this variation may not be linear. They can evoke action potentials in postsynaptic fibers at a frequency that is related to their size. Thus, a generator potential *modulates* the frequency of a sequence of action potentials in a way that reflects the intensity of the stimulus. In the squid eye these events all take place in a single receptor cell: light causes an intensity-dependent change in the polarization of the membrane and this generator potential modulates the frequency of a continuous volley of action potentials that are carried through the optic nerve fibers (the fibers being processions of the receptor cells). Such generator potentials have also been detected in other sensory systems. They seem to involve ion translocations across receptor membranes, similar to those found in the cell membrane of the rods and the cones of the visual receptor and the hair cells in the inner ear.

The basis question in sensory transduction is, how does a stimulus such as light, sound, or a chemical substance initiate a change in the electrical polarization of a membrane involving the movement of ions? A few clues to the answer exist for the visual process and the senses in the ear; they are given in the sections 11.2. and 11.3. The transduction process in the other sensory receptor organs still is poorly understood.

11.2 The visual receptor

Light-sensitive receptors

Light-sensitive receptors of various forms, ranging from very primitive systems to very elaborate ones, are found in most living organisms. The most primitive forms include receptors for phototaxis and photokinesis (light-induces motion) in plants and some algae and also the light-sensitive cells or clusters of cells found on the surface of worms and molluscs. More highly developed animals possess the light-sensitive organs known as eyes. In eyes an image of the enviroment is formed on a "screen" made up of light-sensitive cells. The image formation can be simple like that in the pinhole eyes of some simple marine

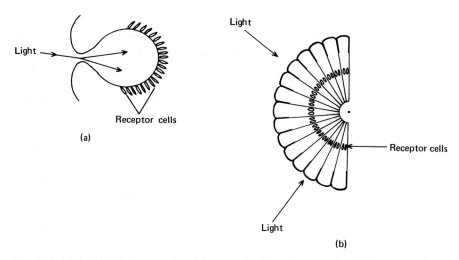

Fig. 11.3. (*a*) A "pinhole" eye, as found in some simple marine animals. (*b*) A compound eye, found in insects and spiders.

animals (Fig. 11.3a). Because of the small aperture, relatively little light is admitted and the sensitivity is not very high. The *compound eye* of insects and spiders is more elaborate (Fig. 11.3b). Compound eyes consist of many separate channels, called *rhabdomeres*, which give the eye its faceted appearance. Light coming from a certain direction can enter only those rhabdomeres that point in that direction. Finally, there is the vertebrate eye which is very much like a conventional camera; an image of the environment is formed by a system of lenses on a screen called the retina.

Figure 11.4 is a schematic representation of a horizontal cross section of a vertebrate eye. The lens system has three components: first, there is the

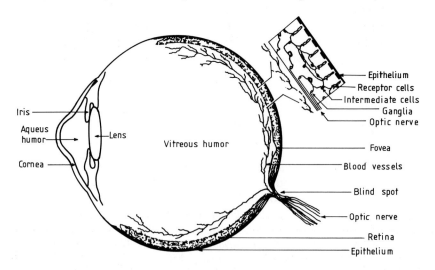

Fig. 11.4. A schematic diagram of the human eye.

cornea with the aqueous humor behind it; second, the lens itself which has a convexity that can be changed by the *ciliary muscle*, thus providing an adjustable focus; and finally, the vitreous humor in the eyeball. This system projects an image of the environment onto a layer containing the receptor cells in the back of the eye; this layer is the *retina*. The amount of light entering the eye is automatically controlled by a circular muscle called the *iris*.

The vertebrate retina

The receptor cells in the retina are embedded in a pigmented epithelium layer and are covered by a layer of intermediary cells, nerve fibers, and blood vessels (see also Fig. 11.1). Thus, light must pass through a layer of tissue before it reaches the receptors. The evolution of the vertebrate eye has tolerated this imperfection apparently because, so far, it does not have much selective disadvantage. The epithelium bed contains a black pigment which serves to reduce the haze of scattered light. The central part of the retina is called the *fovea*; it is here that acuity of vision is sharpest. The spot at which the optic nerve fibers pass out of the eye is a "blind spot" because it cannot contain receptor cells.

There are two kinds of receptor cells in the vertebrate eye (see Fig. 11.5). The *rods* ar elongated cylindrical cells with a narrow "waist" separating an outer and an inner segment. In the outer segment are found a few thousand separate thylakoids, the flat membrane-bound sacs which we have also seen in the grana of the green leaves of plants. These thylakoids contain the visual pigment. The inner segment contains mitochondria and the nucleus; the synaptic

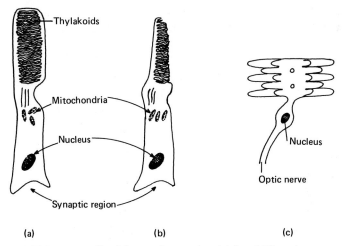

(a)   (b)   (c)

Fig. 11.5. (*a*) and (*b*) Receptor cells of the vertebrate retina. (*a*) A rod. The outer segment contains several hundred thylakoids, each about 200 Å thick, that contain the visual pigment. The length of the outer segment is about 10 μm in man. (*b*) A cone. The cytoplasmic membrane, folded-in repeatedly, contains the visual pigment of the cone. (*c*) A photoreceptor cell of a squid. Thousands of tiny outgrowths (only a few are shown) are sticking out of a cylindrical cell. The cell is a specialized neuron; the axons of many photoreceptor cells form the optic nerve.

regions are at the far end of the cell membrane. The cones have a conical outer segment. The visual pigment in the cones is also confined to membranes, but not in separate thylakoids as in the rod. It is contained in the cell membrane itself which is folded repeatedly at one side, forming a stacked structure somewhat similar to the thylakoid stacks in the rod.

Cones are more concentrated in the central part of the retina and rods are more abundant in the periphery. Cones are specialized for vision in bright light and for color vision; rods for vision in weak light. Rods can be extremely sensitive to light; they respond already to the absorption of one photon. However, their sensitivity depends on the light intensity; at increasing intensity they become less sensitive (become "saturated") until at a certain intensity rod vision changes to cone vision. The acuity in the fovea is a result of the fact that it contains only cones. Cones are less sensitive than rods but since for the most part each cone is connected, through a bipolar cell, to a separate ganglion cell, they make for high acuity; stimulation of two adjacent cones can lead to the firing of separate fibers in the optic nerve.

In more primitive eyes the receptor apparatus is far less organized. In the squid eye, for example, the receptor cells have thousands of little processions called microvilli, which contain the visual pigment (Fig. 11.5c). These cells are actually neurons with axons forming the optic nerve.

Visual pigments

The visual pigment found in rods of most vertebrates is *rhodopsin*, a compound consisting of a protein called opsin and a chromophore which is responsible for the absorption of light in the visible spectral region. The chromophore is *retinal*, a chain of conjugated double bonds (thus an extensive π system) attached to an ionine ring at one end and an aldehyde group at the other. Retinal is a derivate of vitamin A obtained by oxidation of the alcohol residue to aldehyde (see Fig. 11.6). *In vivo* the oxidation is coupled to the reduction of NAD^+ to NADH. Although it is the most abundant, rhodopsin is not the only visual pigment found in nature. There are more visual pigments which are all complexes of a specific opsin and retinal. The opsin forms an

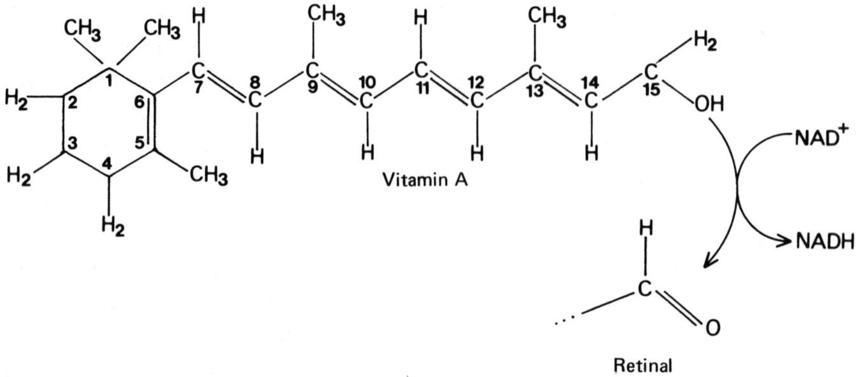

Fig. 11.6. The chemical structure of vitamin A, which is oxidized to retinal by NAD^+.

262

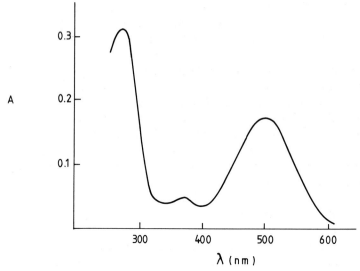

Fig. 11.7. The absorption spectrum of rhodopsin.

environment which determines the absorption spectrum of the retinal. Thus, different cones contain pigments which have absorption spectra which are not only different from that of rhodopsin in rods, but also from each other. Color vision is made possible that way (see below).

Retinal is found in two forms, corresponding to two forms of vitamin A. The most common form is $retinal_1$ (R_1) which is part of rhodopsin as well as most cone pigments. The other form, R_2, differs from R_1 in that its π-systems is extended to carbons 3 and 4 in the ionine ring. It is found in the rods and cones of many fish and of tadpole.

The absorption spectrum of rhodopsin is given in Fig. 11.7. The main absorption band in the visible spectral region peaks at about 500 nm which is also the most effective wavelength for human vision in a dark-adapted eye exposed to weak light. In strong light the peak wavelength of such a visibility curve is shifted to about 550 nm (which is close to the absorption maxima of the cone pigments). The agreement of the visibility curves with the absorption spectra provides evidence that in weak light, when rod vision is predominant, rhodopsin is responsible for vision while in strong light, when rod vision changes to cone vision, the cone pigments are involved.

The amino acid sequences of rhodopsin from several sources have been determined. All these sequences show a high degree of homology and reveal regions of predominantly polar and charged amino acids which are separated by seven stretches of 21 to 28 predominantly nonpolar amino acids. A model based on a number of topographic studies is given in Fig. 11.8. It shows rhodoposin as a transmembrane protein with seven hydrophobic α helices passing back and forth through the thylakoid membrane and linked together by loops sticking out in the cytoplasmic and thylakoid inner space. The retinaldehyde chromophore is buried deep inside the molecule. The complex

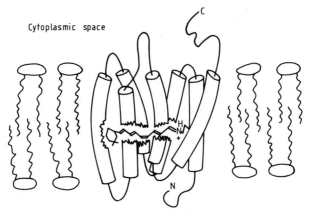

Thylakoïd inner space

Fig. 11.8. A schematic representation of the structure of rhodopsin in the thylakoid membrane of a vertebrate rod receptor cell. The cylinders represent α helices, which are embedded in the membrane. The retinal is buried deep inside the molecule.

is very much alike the protein complex that spans the purple membrane of the bacterium *Halobacterium halobium*. This protein, which is called *bacteriorhodopsin*, is involved in the light-induced transport of protons across the bacterial membrane, a process that converts the electromagnetic energy of the light into chemical energy (see Chapter 8). Although in both cases light

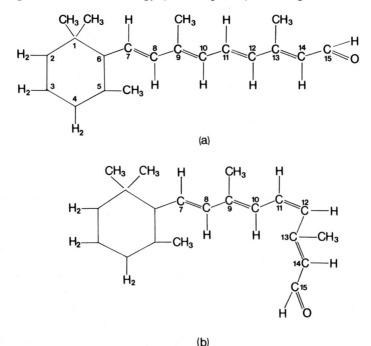

Fig. 11.9. The two isomeric forms of retinal. (*a*) The all-trans form; (*b*) the 11-cis form.

264

causes an isomerization of the chromophore (see below) there are no indications that further steps in the processes are similar.

Light causes the bleaching of rhodopsin in both the intact eye and *in vitro*, in excised retinas as well as in solution. Recovery of the absorption occurs if the retina is intact and metabolically active. The bleaching occurs through a detachment of the retinal from the opsin. The recovery involves the reduction of retinal to vitamin A, the enzymatic isomerization and the reconversion of the vitamin A, and the reattachment of the chromophore to the protein. Due to the pioneering results from the group of George Wald it is now known how this is accomplished.

Retinal, and also vitamin A, can exist in various stereoisomeric forms. The 11-cis isomer and the all-trans isomer are important for vision (Fig. 11.9). These stereoisomers are interconvertible by a photochemical reaction. The 11-cis isomer spontaneously combines with opsin; the all-trans isomer does not. Thus, in rhodopsin the chromophore is bound to the protein in its 11-cis configuration. This bond is by a so-called *Schiff-base* linkage; it is formed when the aldehyde group of the retinal reacts with the amino group of an amino acid (lysine) of the opsin, expelling a water molecule:

$$\underset{\displaystyle |}{\overset{\displaystyle H}{}} \qquad\qquad \underset{\displaystyle |}{\overset{\displaystyle H}{}}$$

$$\text{Ret}-\text{C}=\text{O} + \text{H}_2\text{N}-\text{lys}-\text{Opsin} \rightarrow \text{Ret}-\text{C}=\text{N}-\text{lys}-\text{Opsin} + \text{H}_2\text{O}. \qquad (11.1)$$

Photochemistry of rhodopsin

When rhodopsin is excited, the 11-cis retinal converts to all-trans retinal. The shape of the chromophore is now incompatible with the opsin and the protein probably goes through a number of configuration changes before the all-trans retinal comes off. It is then reduced to all-trans vitamin A and reconverted either by light or enzymatically (involving the enzyme isomerase) into the 11-cis form. The 11-cis vitamin A is then oxidized to 11-cis retinal and linked by a Schiff-base linkage to the opsin back in the thylakoid membrane. The protein then assumes its original conformation characteristic of rho⁻ dopsin.

The intermediate stages which we have assumed to be different conformations of the opsin can be followed by spectrophotometric experiments at different temperatures; except for the first step, the photoisomerization proper, all successive steps are temperature dependent. If a suspension of rhodopsin is brought to liquid nitrogen temperature ($77°K = 196°C$) and excited by a flash of light its absorption spectrum changes. Some of the rhodopsin with an absorption maximum at 500 nm will disappear and a substance with a maximum absorption at 543 nm will be formed. This first product of the photochemical reaction sequence is all-trans retinal bound to opsin, called *prelumirhodopsin*. Warming up the suspension gradually allows us to follow the subsequent steps in the series. The next substance formed, appearing at temperatures above 140°C, is *lumirhodopsin*. We then find *metarhodopsin I*, above 40°C, and *metarhodopsin II*, above 15°C. Metarhodopsin II decays either to *pararhodopsin* or to all-trans retinal and opsin at temperatures above 0°C. Pararhodopsin itself also decays into free all-trans retinal and opsin. Because of the photochemical interconvertibility of the isomers of retinal, light can reconvert every intermediate in the sequence back to rhodopsin. This sequence of events is shown schematically in Fig. 11.10.

The photointerconvertibility of the chromosphores causes the phenomenon of the photostationary state. Suppose a solution of rhodopsin at liquid nitrogen temperature is exposed to a brief flash of light, so that many molecules of rhodopsin are converted to prelumirhodopsin. A second flash of light will excite more rhodopsin and also some prelumirhodopsin formed during the first flash. The excited molecules of both thypes, can thus decay partly to one isomeric form and partly to the other; that is, rhodopsin to prelumi-rhodopsin and prelumirhodopsin to rhodopsin. Consequently, under conti-nuous illumination a steady state can be established in which the rates of conversion of one pigment to the other are equal. The steady-state condition is called the photostationary state; the relative proportions of the two pigments in this state are determined by the relative rates at which the pigments absorp light (their absorption coefficients and their concentrations) and the relative probability that each kind will decay to its photochemical product (rhodopsin for prelumirhodopsin and prelumirhodopsin for rhodopsin). If the absorption

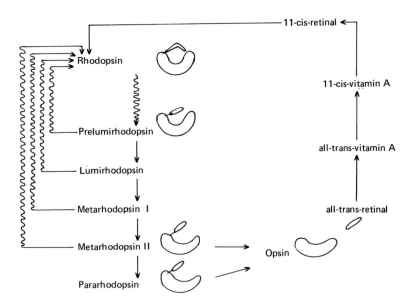

Fig. 11.10. A diagram of the photochemistry of vision. Light isomerizes 11-cis retinal to all-trans retinal in the first intermediate product prelumirhodopsin. The protein then goes through a number of intermediate configurations until the all-trans retinal comes off. After oxidation, isomerization into the 11-cis form and reduction, rhodopsin is again formed. Since the pho-toisomerization is reversible, as long as the retinal stays on the opsin, rhodopsin can be reproduced by light (designated by the wavy arrows).

266

spectra of both pigments are different (which is true in the case of rhodopsin and prelumirhodopsin), one can let the concentration of one predominate over the other by a suitable choice of the wavelength of the exciting light. Light of 600 nm, which is absorbed far more by prelumirhodopsin than by rhodopsin, will cause the concentration of rhodopsin to predominate over that of prelumirhodopsin in the photostationary state. Conversely, the use of 450 nm light, absorbed more strongly by rhodopsin, causes a higher concentration of prelumirhodopsin in the photostationary state. Photostationary states are set *in vivo* and are also important, for instance, in the setting of light-influenced daily circadian and other periodic rhythms.

ERG and ERP

The sequence of steps in the photochemical cycle of rhodopsin can also be followed by measuring electrical potential changes caused by light entering an intact eye. If an electrode is placed on the cornea and another behind the retina (of in the mouth) a potential change will occur between them when light is allowed to enter the eye. A recording of such changes is known as the *electroretinogram* (ERG). The changes start a few milliseconds after the onset of illumination and show typical waves labeled a, b, c and d in Fig. 11.11. These potential waves apparently reflect changes in the electrical activity

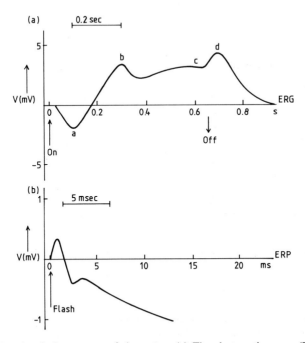

Fig. 11.11. The electrical responses of the retina. (*a*) The electroretinogram (ERG) occurring a few milliseconds after the onset of illumination, showing the a, b, c, and d waves. (*b*) The early receptor potential (ERP) occurring immediately after flash illumination, showing a positive peak followed by a negative one.

267

(mobility of ions) of the membranes of the receptor cells and interneurons. When strong but short flashes are used instead of continuous illumination, a smaller electrical effect can be detected that begins with no detectable lag period and lasts a few millisecond, thus filling the initial gap of the ERG. This more rapid response is called the *early receptor potential* (ERP) or *fast photovoltage* (FPV) and most probably reflects the formation of electric dipoles attending the photochemical steps. It shows a positive peak, labeled R_1, immediately followed by a negative peak R_2, after which the ERG starts to develop. At temperatures below –15°C the R_2 peak disappears. It has been established that the rise of R_1 is associated with the formation of metarhodopsin I to metarhodopsin II. By using closely spaced pairs of flashes the conversions of the intermediates of the cycle back to rhodopsin can be followed, and the amounts of the intermediates formed together with the kinetics of their formation can be revealed. Experimentation of this kind has contributed substantially to our knowledge about the photochemical cycle.

The visual generator potential
Information about the way in which the photochemistry of rhodopsin causes generator potentials, which then give rise to the firing of action potentials in optic nerve fibers, has been derived from biochemical determinations and from experiments in which microelectrodes are implanted in selected places so as to study electrical activities of single cells or small groups of cells during stimulation by light. In vertebrates, most detailed work is done on rods, particularly those of amphibians, which are largest and therefore most accessible. From such work it has been established that, unlike most other excitable cells, which depolarize on stimulation (see Chapter 9), vertebrate rods *hyperpolarize* from a resting potential of about –40 mV in the dark to about –80 mV when illuminated. The resting potential in a rod is kept at its "dark" level by an *inward current of Na$^+$ ions*, a behavior that also deviates from that of other excitable cells. Upon stimulation, the Na$^+$ current is blocked, thus causing the hyperpolarization. The hyperpolarization is communicated to the horizontal cells, but in the bipolar cells it may be either a hyperpolarization or a depolarization. This seems to be related to the patterns of antagonism between central and peripheral regions of a receptive field (see section 11.1). The amacrine cells are all depolarized and occasionally show an action potential. The ganglion cells respond with numerous action potentials and only a slight depolarization.

The hyperpolarization of the plasma membrane of a rod cell is caused by the photoisomerization of retinal, buried in rhodopsin which sits in the thylakoid membrane. It is clear, therefore, that somehow the signal must be transmitted from the thykaloid to the plasma membrane. How is this accomplished and what is the transmitter? It has been thought for some time that Ca^{2+} ions, which do have a significant role in vision, carry out this function. In such a model, excitation of rhodopsin would cause the release of Ca^{2+} from the thykaloid inner space into the cytoplasm. The Ca^{2+} then would block the Na$^+$ channels in the outer membrane thus causing hyperpolarization. An avalanche of recent evidence, however, points to *cyclic guanine monophosphate* (*c*GMP) as being the transmitter. *c*GMP is a guanine nucleotide in which a phosphate group joins the 3′ and the 5′ carbons in the ribose together, thus giving it a ring structure. The level of *c*GMP in rod cells is unusually high. It is hydrolysed in response to light. An enzyme, called *c*GDP-specific *phosphodiesterase* (PDE) seems to be responsible for this rapid hydrolysis. Electrophysiological measurements have shown that intracellular injection of *c*GMP mimics the dark state in the rod cell. It appears, therefore, that high levels of *c*GMP keep the Na^+ channels open and that hydrolysis of *c*GMP, a reaction that breaks the ring structure, is involved in closing them. The hydrolysis occurs when the enzyme PDE is activated by light.

How do these events follow from the bleaching of rhodopsin? An answer to this question became possible when it was discovered that bleached rhodopsin was able to bind a specific protein that recently was isolated by the group of L. Stryer. This protein was given the name *transducin*. In dark-adapted rods transducin consists of three subunits, designated as α, β, γ. The α subunit is bound to guanine diphosphate (GDP). The complex is a peripheral protein and can be visualized on the thykaloid surface as small particles. Upon illumination, the complex binds to bleached rhodopsin. When this happens, the complex exchanges the bound GDP for a GTP (guanine triphosphate) and the α subunit containing GTP is released from the membrane, separately from the other two subunits. The free α-transducin·GTP complex then binds to PDE, activating it and thereby initiating the hydrolysis of *c*GMP. The activation of PDE apparently is a result of the removal of an inhibitory subunit from the enzyme.

The mechanism that we just described is also able to explain the extreme sensitivity of a rod cell. It contains two cascading steps. The first one occurs when transducin is activated by binding to the photobleached rhodopsin. One single photobleached rhodopsin molecule can release hundreds of α-transducin·GTP complexes. The second step is the hydrolysis of some 4,200 molecules of *c*GMP per second by each activated PDE molecule.

Light also stimulates somehow the activity of *guanyl cyclase*, an enzyme that catalyzes resynthesis of *c*GMP. The result of this is a light-initiated increase in *c*GMP turnover. Since, apparently, calcium ions can modulate guanyl cyclase activity, thus regulating the level of *c*GMP, calcium may be important for adaptation of the eye to different light intensities. The evidence for such a role of calcium ions still is scarce, however.

Color vision
Color vision seems to be possible when two different wavelengths of light cause relative stimulation in two or more types of receptor cells. Although two kinds of receptors with different absorption spectra are enough to provide some sense of color, the existence of three different types is suggested by J.C. Maxwell's demonstration that nearly all colors can be reproduced by mixing of three "primary" colors chosen in such a way that none of the three can be made by mixing the other two. Measurements of spectral sensitivities in

cones of certain monkeys (which are thought to have photoreceptors like those in man) have identified three groups of pigments with maximum absorption at 430, 530, and 580 nm. Further evidence indicated that each cone contains only one type of pigment, so that there are in fact different groups of cones with overlapping but distinct absorption spectra.

It seems, however, that the perception of color requires a good deal of processing in the interaction neurons as well as in the brain itself. The experiments reported in the late 1950s by E. Land gave quite unexpected results considering three-color theories as described above. A striking example is the experiment in which a reproduction of a colorful object, such as a basket of fruits and flowers, was produced by superimposing projections of two photographic slides of the same object. One, uncolored, photograph was taken through a color filter that transmits only light in the red part of the visible spectrum and the other, also uncolored, through a filter transmitting yellow-green light. When the slide produced in red light was projected, using red light, on top of a projection with white light of the other slide, the full gamut of colors appeared on the screen. The colors perceived in the experiment were largely independent of the intensities of the projection lamps.

Land described this and later, similar, results by a mathematical model which he calls the Retinex Theory. In this model the color of a point of the receptive field can be predicted by taking computing relationships between the radiation from that point and from all other points of the receptive field.

11.3 The otic receptors

The hair cell
Hearing, like vision, is a major sensory system. It enables many animals, including humans, to perceive their environment through mechanical vibrations within given frequency ranges. Although the study of hearing is one of the oldest fields in biophysics, little is known about its transduction mechanism proper. The reason is most probably the poor accessibility of the system to experimental approach. To obtain relevant information the system must be pretty much intact, and under such conditions it is extremely difficult to reach the inner ear, where the transduction organ is located. The system is enclosed in deep cavities of the skull; to reach the inner ear, parts of the temporal bone have to be cut away, which is difficult to do without extensively damaging the system. Anatomical and histological studies have revealed much about the structure of the ear, but not many clues to the mechanism of transduction. The early biophysical studies were predominantly limited to measurements of impedance and impedance matching devices of the sonic energy transmission system.

Recently, however, data became available about the working of a crucial cell of this system, the *hair cell*. Aquiring such data became possible with the realization that the inner ear does not contain only one but in fact *six*

270

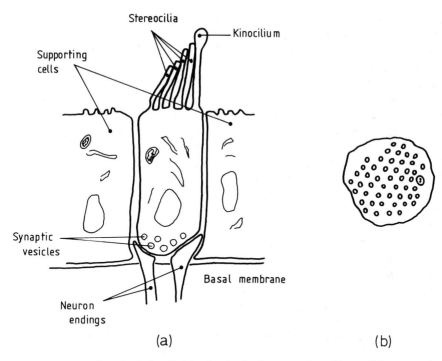

Stereocilia

Kinocilium

Supporting cells

Synaptic vesicles

Basal membrane

Neuron endings

(a)

(b)

Fig. 11.12. A drawing of a hair cell; (*a*) a longitudinal cross section; (*b*) a radial cross section across the apical surface.

sensory systems which all operate according to the same principles. In addition to hearing proper there are the three senses that detect rotations about three independent axes and the two senses that detect acceleration in horizontal and vertical directions. All these senses involve transduction of mechanical energy into a (change in) membrane potential, accomplished by hair cells.

Hair cells are named for the bundle of hair-like projections that extend from their upper surface. Figure 11.12 is a schematic drawing of a hair cell together with supporting cells forming an epithelical sheet. At the apical end of the cell there is a bundle of extensions which lean together to form an obliquely truncated cone. The extensions form an hexagonal array sticking out from the apical surface (Fig. 11.12b). One of the extensions is somewhat different from the others. This is the *kinocilium*. It has an internal structure similar to cilia and flagella (see Fig. 10.1) but is unable to perform motion. The other extensions are called *stereocilia* although they are not like cilia at all. Their internal structure is formed by actin fillaments. The cilia are arranged in a highly regular manner. The bundle is approximately circular in cross section and along one diameter of the circle there is a progressive increase in the length of the stereocilia. The cilia along any axis perpendicular to that diameter are equal in length, however, the bundle thus has a bilateral plane of symmetry.

At the basal surface of the cell are synaptic couplings with two types of

271

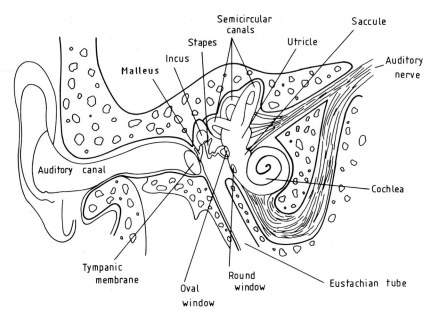

Fig. 11.13. A drawing of a cut-open human ear showing the outer, middle, and inner ears. The outer ear is the auricle and the auditory canal; the middle ear is the air-filled space between the tympanic membrane and the oval window, containing the middle ear ossicles; the inner ear is the coiled tubular structure.

neuron endings, one sending messages to the brain (afferent ending), the other receiving messages from the brain (efferent ending). The significance of the latter is not quite clear. Near the base of the cell are vesicles containing neurotransmitter. The composition of the neurotransmitter is not known. Hair cells are very sensitive transducers; a displacement of a few Ångstroms already evokes a response in the form of a change in membrane potential.

The anatomy of the ear
Figure 11.13 is a sketch of the human ear. It consists of three parts. First, there is the outer ear with the *pinna* or *auricle* and the *outer auditory canal*. At the end of the auditory canal is the *tympanic membrane*, or ear drum, which separates the outer ear from the middle ear cavity. In this air-filled space are three tiny bones called the *middle ear ossicles*; these are the *malleus* (hammer), the *incus* (anvil) and the *stapes* (stirrup). The outermost of the three is the malleus which is pressed against the tympanic membrane. The innermost of the bones, the stapes, pushes against a membrane called the *oval window* which separates the air-filled middle ear cavity from the liquid-filled channels of the *inner ear*.

When mechanical vibrations of air enter the outer auditory canal the tympanic membrane is set in motion. This motion is transmitted to the middle ear ossicles which together act as an amplifier system. The bones, in fact, form a lever system that decreases the amplitude of the vibrations of the tympanic system but increases the force exerted on the oval window, thus performing an impedance matching between the tympanic membrane and the fluid in the cochlear channels. The amplification amounts to about a 25 db gain in acoustic pressure. Both the malleus and the stapes have muscles which, upon contraction, can change the elasticity of the system, thus providing a volume control.

A narrow channel called the *Eustachian tube* connects the air-filled space of the middle ear with the pharynx. This connection is necessary in order to maintain equal average pressure on both sides of the tympanic membrane.

The channels of the innner ear form a structure (sometimes referred to as the *labyrinth*) consisting of six major systems. The *semicircular canals* are the three tubes on top of the structure, bent in the form of a circle, which detect angular acceleration about three independent axes. Underneath of them are the *utricle* and the *saccule* which detect linear acceleration each in a different direction. The coiled structure at the bottom is the *cochlea* that contains the hearing detectors. Each of these six systems can be seen as a pouch of epithelium. Inside the pouches is a fluid called *endolymph*; outside the pouches, between the epithelium and the bone, is a different fluid called *perilymph*.

The semicircular canals
Angular acceleration is sensed by the three semicircular canals. The sense of balance that enables humans to walk upright is made possible by this system together with the systems located in the utricle and the saccule, that detect linear acceleration. To this purpose, the three semicircular canals are joined together in such a way that when the head is in its usual orientation one canal lies in de *XY* plane of a three dimensional space, one in the *XZ* plane and one in the *YZ* plane. In a part of each canal the cross-sectional area is somewhat larger than in the rest of the canal. This part is the *ampulla*

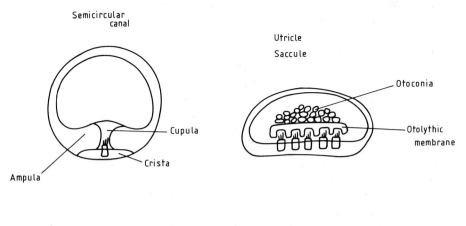

(a)   (b)

Fig. 11.14. A schematic cross sectional drawing of a. a semicircular canal near its ampulla; b. an utricle or a saccule.

(Fig.11.14a). In it is a septum that contains hair cells. The septum consists of a layer of cells called *crista* with a sheet of extracellular material, called *cupula* extending from it. The hair cells are situated in this layer. The hair bundles are inserted in an upward direction in the gelatinous cupula. When the head is rotated quickly, the inertia of the endolymph in the canals causes it to lag behind the walls of the canal which moves with the head. The fluid then presses on the cupula which, in turn, causes the hair bundles to bend, thus evoking a response of the membrane potential of the hair cells.

The utricle and the saccule

The utricle and the saccule are the organs that detect linear acceleration, the former mainly in the horizontal and the latter mainly in the vertical direction. The hair cells of both are found in crescent-shaped patches of epithelium forming a more or less flat sheet (Fig. 11.14b). In the utricle the epithelical sheet is mainly horizontal and in the saccule it is mainly vertical. Parallel to the epithelical sheet and quite close to it is a structure called *otolythic membrane* which is made up of protein molecules. The kinocilium of each hair cell is inserted into indentations in the otolythic membrane. Piled on top of the otolythic membrane are numerous tiny cristals made of calcite (a form of calcium carbonate) which are called *otoconia*. The density of the otoconia is several times greater than the endolymph inside the utricle and saccule. Therefore when the system is accelerated, the greater inertia of the otoconia causes them to lag behind the endolymph. The lagging motion is communicated to the otolythic membrane which, as a result, exerts a force on the hair cells which is directed against the accelerated motion. This force again is transduced into a change in membrane potential.

The cochlea

The sense of hearing is located in the cochlea. Hearing is the perception of pressure waves that are propagated by a vibrating medium. The waves of vibration enter the hearing apparatus which in alle vertebrates is very homologous to the human ear, although frequency characteristics may vary appreciably from species to species. For the human ear the frequency range is between about 20 Hz and about 20 kHz; the exact limits depend on the individual and are influenced by age and environment.

The cochlea is a tube that is coiled into a helix of two and a half turns, decreasing in size as it turns away from the middle ear. Two longitudinal membranes, divide the "tube" into three ducts (Fig. 11.15) all filled with fluid. Below the *basilar membrane* is the *tympanic channel* which is filled with perilymph. Above the basilar membrane are two ducts separated by *Reisner's (vestibular) membrane*. These ducts are the *vestibular channel* and the *cochlear channel*. The vestibular channel connects with the tympanic channel through an opening called the *helicotrema* and, thus, also contains perilymph. The fluid in the cochlear channel is endolymph. One could visualize the three ducts as being formed by a long tube which is folded, thus forming two interconnected ducts with a third one pressed in between (see Fig. 11.15b).

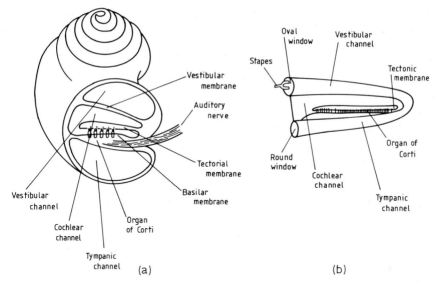

Fig. 11.15. A crosssectional view of the cochlea (a) and a schematic representation of the organization of the three liquid filled channels in the cochlea.

The oval window is at one end of the vestibular channel. The footplate of the stapes fits against this window; vibrations picked up by the middle ear ossicles are transmitted to the perilymph across the oval window. The end of the tympanic channel is the round window.

Seated on the basilar membrane is the *organ of Corti*. The hair cells are in this organ. The hair bundles are inserted in a structure called the *tectorial membrane* that is placed parallel to the basilar membrane and very close to it. Since the basilar membrane and the tectorial membrane are "hinged" along different axes, the upward displacement of the membranes caused by the reaction of the stapes to sound waves is accompanied by shearing motions between them. These relative motions bend the hair bundels extending between the two membranes, thus causing changes of the membrane potential in the hair cells.

The operation of the cochlea is very complicated and not yet fully understood. It still is a matter of considerable controversy. Some pertinent information has been obtained by observing the responses of the basilar membrane to sound of different frequencies and by using microelectrodes placed in the cochlear spaces and in fibers of the auditory nerve. The movement of the basilar membrane as a response to sound is actually a traveling wave having an amplitude maximum at a distance from the oval window which is a function of the frequency of the exciting sound wave. Some typical frequency responses are shown in Fig. 11.16.

The transduction mechanism

Although the details of the transduction mechanism still are not fully understood, recent experiments with single hair cells have given some indi-

275

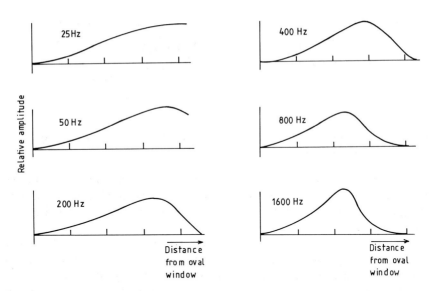

Fig. 11.16. Displacements of the basilar membrane as a response to sound of different frequencies (from von Békèsy (1953), *J. Acoust. Soc.* **25**, 786).

cations how the transduction might work. The resting potential of a hair cell, most probably maintained in a manner similar to that in which it is in neurons, is about -60 mV. Measurements on single hair cells from the saccule of the American bullfrogg have revealed that the membrane potential is changed upon movement of the hair bundle. The response is direction-dependent. When the tip of the hair bundle is moved along the axis of bilateral symmetry *towards* the kinocillium a depolarization to about -40 mV occurs. If the hair bundle is pushed along the same axis but *away* from the kinocilium a (small) hyperpolarization is seen. The sensitivity decreases strongly in all other directions. Very small displacements in the sensitive direction (in mammalian hair cells around 1 Å) already evokes a response.

The depolarization, respectively hyperpolarization most probably is due to an influx, respectively efflux of positive ions. The channels do not seem very specific to a special type of ion. However, a major involvement of potassium ions in the charge movements seems likely because of the high levels of this ion in the endolymph. A responsive transport of calcium ions through calcium selective channels near the basal area of the hair cell may cause the release of neurotransmitter and consequent synaptic transmission.

11.4 The chemical, somatic, and visceral receptors

Chemical stimulation
The chemical receptors are those of olfaction and taste. Although they both respond to chemical stimulation, they are quite different otherwise, involving

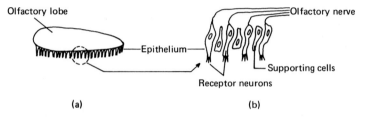

Fig. 11.17. (a) A drawing of the human olfactory receptor (b) a part of the olfactory epithelium showing the receptor neurons.

different sensory pathways to different areas of the brain. Olfaction requires that the stimulating chemical substance be volatile and soluble in the mucous secretions of the upper membranes of the nose. Taste requires that the chemical be soluble in water.

The olfactory receptor is located in the mucous membrane of the upper part of the nasal passage. The receptor cells are part of the so-called *olfactory epithelium* and are, in fact, specialized nerve cells; they have fibers which together form the olfactory nerve (Fig. 11.17) at one end and have little hairy outgrowths at the other end. There is no information about the reactions which evoke action potentials in the olfactory nerve fibers.

The receptor for the sense of taste is the *taste bud* (Fig. 11.18). There seem to be four fundamental sensations of taste each of which has a separate kind of taste bud located on different parts of the tongue. Taste buds for the sensation of sweetness and saltiness are mostly located at the tip of the tongue. Sourness is detected by taste buds distributed along the sides of the tongue and the back of the tongue contains the buds for the taste of bitterness. All taste buds have the ovoid structure shown in Figure 11.18 but they can differ substantially in sensitivity and specificity of taste. Furthermore, the receptor cells in taste buds are specialized neurons; their outgrowths at one end are axons together forming the gustatory nerve.

Somatic stimulation

Somatic receptors can be distinguished as those responsible for touch and pressure, those for heat and cold, and those for pain. The sensation of touch seems to be provided by three types of receptors. The bending of a hair can cause a single discharge from a myelinated fiber within an area of 1.5-5 cm². Continuous discharge can be maintained by another type of receptor under

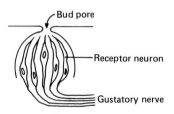

Fig. 11.18. A drawing of the human taste bud showing the receptor neurons.

277

pressure. The "field" for such mechanoreceptors is very small, something like 1-2 mm in diameter. The rate of the discharges can be very high (330 discharges/ sec under high pressure). Finally, there are receptors which are sensitive to both pressure and cooling. The rate of discharge from these receptors is usually much smaller than that of the pure pressure-sensitive type.

Two different sets of receptors are responsible for the sensation of temperature, one initiating firing more rapidly when the temperature of the skin drops and the other more rapidly when the temperature is raised. The skin seems to be divided into spots that are sensitive only to cooling and spots that are sensitive only to heating. There is little information about the anatomical, let alone the physiological, character of these receptors.

Pain receptors do not have the high specificity that other receptors have; they are spread throughout the body. Electrical and mechanical disturbances, extremes in temperature, and chemical reactions can all initiate pain responses. Although the sensations of pain, touch and pressure, and temperature are different and thus most involve different types of receptors, they quite often go together. Moreover, the firing of the nerve fibers is also initiated directly by touch, cold, and heat or by combinations of these stimulants. This is the reason why it is so extremely difficult to identify the individual receptors.

Visceral sensory signals
Very little can be said about the mechanisms of visceral sensations. For most of them it is apparent that they originate in the visceral (internal) organs (hunger, for example, may be initiated by rhythmical contraction of the muscles in the wall of the stomach; thirst is related to a reduction of the body fluid content). The mechanisms by which all these sensations are evoked are totally unknown.

From a biophysical point of view the study of the "lower" sensory systems has not been very rewarding. This is largely due to the fact that little is known about the anatomy and histology of these systems and hardly anything about their physiology. The approach to these systems is extremely difficult and it may very well be that more understanding about their function can only be expected from an increased knowledge of the functions of the central nervous system.

Bibliography

Dallos, P. (1973). "The Auditory Periphery: Biophysics and Physiology". Acad. press, New York.
Hargrave, P.A. (1986). Molecular Dynamics of the Rod Cell, in "The Retina. A Model for Cell Biology Studies" (R. Adler and D. Farber, Eds.), pp. 207–237, Acad. Press, Orlando Fl.
Hudspeth, A.J. (1983). The Hair Cells of the Inner Ear, *Sci. Am.* **248** (January), 42–52.
Land, H.E. (1959). Experiments in Color Vision, *Sci. Am.* **200** (May), 84–99.
Land, E.H. (1983). Recent Advances in Retinex Theory and Some Implications for Cortical Computation: Color Vision and Natural Image, *Proc. Natl. Acad. Sci. U.S.* **80**, 5163–5169.
Levine, J.S., and MacNichol Jr., E.F. (1982). Color Vision in Fishes, *Sci. Am.* **246** (February), 108–117.

Schnapf, J.L., and Baylor, D.A. (1987). How Photoreceptor Cells Respond to Light, *Sci. Am.* **256** (April), 32–39.

Stryer, L. (1987). The Molecules of Visual Excitation, *Sci. Am.* **257** (July), 32–40.

Von Békèsy, G. (1953). Shearing Microphonics Produced by Vibrations near the Inner and Outer Hair Cells, *J. Acoust. Soc. Am.* **25**, 786.

Wald, G. (1955). The Photoreceptor Process in Vision, *Am. J. Ophthalmol.* **40**, 18–41.

Wald, G. (1968). The Molecular Basis of Visual Excitation, *Science* **162**, 230–239.

12. Theoretical biology

12.1 Physical concepts and biology

Theoretical biology is a subdiscipline of biology that may have a significant impact on the further development of biology. Its meaning for the biological sciences is comparable to the meaning of theoretical physics for the physical sciences. Theoretical physics, and especially mathematical physics, is concerned with the description of physical phenomena in mathematical terms. Beyond this, however, it makes abstractions from physical reality and puts them into a conceptual framework in such a way that from abstract models predictions can be made about the phenomenology of real systems. Theoretical biology is trying to do just that; and the fact that in doing so many concepts of the physical sciences are "taken over" and used to describe biological phenomena in an abstract manner provides an argument for considering theoretical biology as a legitimate domain in biophysics. Its aim, of course, is the unification of assemblies of facts and their simplification through the logic of a (abstract) concept, thus arriving at a deeper understanding and greater predictability.

In the preceding chapters of this book we have seen physical concepts being used to explain biological phenomena as well as put them into a framework wherein closer investigation becomes more systematic. Examples of this are the uses of quantum mechanics and thermodynamics in our discussions. This is not really theoretical biology, but in some of the subjects discussed (i.e. membrane transport) the application of physical concepts can lead to a much more theoretical treatment. In this chapter we will briefly discuss the way in which some of these physical concepts are used in theoretical biology and to what kind of results this application may lead. The purpose of this discussion is to introduce the reader to these problems, rather than to treat them *in extenso*. Moreover, the choice of topics is somewhat arbitrary and far from complete. Readers who desire to pursue this domain more thoroughly are referred to the literature.

The concepts which have been used in theoretical biology are those of quantum mechanics, statistical mechanics, thermodynamics, and, more recently, cybernetics and information theory. Applications of quantum mechanics involve attempts to deal with the stability of a (living) system operating far from equilibrium and with the collective behavior of a many-unit system (such as an organ consisting of a large number of cells). However, the usefulness of quantum mechanics for the description of such systems is much debated (see, for instance Fröhlich, 1969). The application of (classical) statistical mechanics and statistical thermodynamics sometimes seems easier to justify, but results have so far given only limited information. An example is a study by Oosawa and Higashi (1967) on the polymerization and polymorphism of protein (such as actin, for example), based on principles of statistical thermodynamics. Kornacker (1972) tried to devise a generalization of statistical mechanics to provide a basis for understanding how the microscopic behavior of nonliving atoms can generate the macroscopic appearance of a "living aggregate". A theoretical approach to the *complexity* of biological systems is described in a recently (1985) published collection of three essays on the subject, edited by R. Rosen.

12.2 Nonequilibrium thermodynamics

A recent addition to the methods of theoretical biology is the approach based on *nonequilibrium thermodynamics* or the *thermodynamics of irreversible processes*. This newly emerging branch of thermodynamics (see for instance Prigogine, 1962, or Katchalsky and Curran, 1967) is a generalization which makes possible a quantitative description of biological phenomena such as the transport of matter, nerve conduction, and muscle contraction. The common factor of all such phenomena is that they are essentially nonequilibrium open systems in which irreversible processes are occurring. Without the formalism of nonequilibrium thermodynamics, a quantitative description of these phenomena could only be based on kinetic equations derived from specific models or mechanisms. However, the detailed information required for an adequate kinetic description is not often available or is difficult to obtain. The thermodynamic treatment is independent of such kinetic or statistical models. This method thus provides an enrichment which is at the same time a limitation: while the method offers additional insights into the factors influencing the phenomena in question, it never can lead directly to the mechanisms of the phenomena. By using nonequilibrium thermodynamics one could determine, for instance, whether or not a model is feasible from a physico-chemical standpoint, but one could never arrive at such a model just by using the method.

Thermodynamics deals with macroscopic quantities, such as pressure, temperature, etc. From a macroscopic point of view we can distinguish between two types of structure: *equilibrium structures* and *dissipative structures*. Equilibrium structures are maintained without an exchange of energy or matter. They form the domain of equilibrium or classical thermodynamics. Dissipative structures are maintained *only* through the exchange of energy and in some cases also that of matter. A living system is typically a dissipative structure. Nonequilibrium thermodynamics is an extension (or rather a generalization) of equilibrium thermodynamics which include dissipative structures. Its basis is the splitting of the entropy term (see Appendix II) dS into a part describing the flow of entropy due to interactions with the environment, d_eS and a part describing the production of entropy inside the system, d_iS. Thus,

$$dS = d_eS + d_iS \tag{12.1}$$

or

$$\frac{dS}{dt} = \frac{d_eS}{dt} + \frac{d_iS}{dt}. \tag{12.2}$$

The second law implies that for all physical processes

$$d_iS \geqslant 0. \tag{12.3}$$

Equilibrium thermodynamics deals with equilibrium situations in which the entropy production term d_iS vanishes. In nonequilibrium thermodynamics we study macroscopic states on the basis of their entropy production.

We can define the entropy production per unit time and volume, σ, by

$$\frac{d_iS}{dt} = \int \sigma \, dV \tag{12.4}$$

in which σ is related to flows and forces of the irreversible processes according to

$$\sigma = \sum_i J_i X_i. \tag{12.5}$$

In Eq. 12.5 J_i is a flow, or a reaction rate, of an irreversible process i and X_i is the conjugated generalized force driving process i. If such a process is a chemical reaction, J_i is the reaction rate v and X_i is the chemical affinity A divided by the absolute temperature. $X = A/T$. The chemical affinity is defined as

$$A = \sum_j v_j \mu_j. \tag{12.6}$$

in which v_j is the stoichiometric coefficient and μ_j is the chemical potential of reaction component j.

Elaboration of Eq. 12.5 can lead to a quantitative description of processes like the transport of matter through biological membranes and energy conversion reactions. Interesting implications follow when the theory is extended to nonlinear systems. Prigogine (1969) has shown that in such cases new structures which are the result of instabilities can bring the system to a new steady state which has a higher order of organization than the original state. In physical chemistry and in hydrodynamics there exist a wealth of systems which show this type of behavior. This, evidently would have tremendous implications for the relation between the physical sciences and biology, if it can be shown that biological dissipative structures can arise from such events.

12.3 Modeling

Physical modeling
The application of physical concepts to biology has often led to the design of models. Many different kinds of models of living entities can be created. Although in most cases quantitative information is difficult or impossible to obtain from models, they are very useful from a conceptual point of view, giving insight into the operational strategy of organisms. In this respect they have become an important tool of theoretical biology. We can distinguish physical analog models from mathematical models. A physical analog is a construction that simulates specific functions and/or properties of a real counterpart. A physical analog need not be, but could be constructed in actuality. It could be a model on paper which simplifies the real counterpart in the sense that only the function or property to be studied is represented in the model. A mathematical analysis of the model could then yield information about the real system. An example is a hydrodynamic model of the bloodstream. From such a model useful information can be obtained, for instance, about such matters as pulsate blood flow in elastic tubes, reflections of pulse waves at complex vascular bifurcations, and the effects of viscosity on pulsative flow. Other examples are electrical analogs in which organic functions are simulated by electrical circuits.

An organic function can often be described by the same set of differential equations that describe specific electrical circuits. Such electrical analogs have been made for the cardiovascular system, the myocardium itself, the lung gas flow, and also for things like natural biological (circadian) rhythms. Physical modeling has also become increasingly important in the areas of agronomy and ecology. It should be emphasized, however, that, although such analogs are of considerable interest to physiologists, the models are usually at best very complex but qualitative simulations of the real system. In only very few instances have physical models allowed the prediction of unknown quantitative data.

Mathematical modeling
In physical modeling there is usually some geometric or functional similarity between the model and the real system. In mathematical modeling this similarity is abandoned; in other words, a mathematical model is a much further extended abstraction from the real system. A simple type of mathematical modeling is numerical kinetics simulation. This is often used to obtain data about metabolic processes, to establish time-dependent distribution of substances (drugs, for example) in organs, and to evaluate data obtained from experiments with radioactive or heavy isotopes. The method involves the design of a numerical, so-called compartment model. The compartments in such a model represent pools of the substance under study. Transfer of the substance between the pools is described by rate coefficients. This type of model leads to a set of simultaneous differential equations which can usually be solved

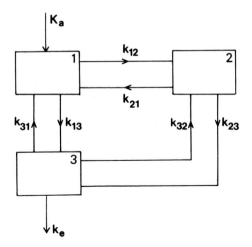

Fig. 12.1. A three-compartment model as described in the text.

after application of a few approximations. A generalized compartment model comprising three compartments is illustrated in Fig. 12.1. If it is assumed that the rate coefficients are constants, the set of linear differential questions describing the time-dependent distribution of a substance N among the three compartments is as follows:

$$
\begin{aligned}
dN_1/dt &= -(k_{12} + k_{13})N_1 + k_{21}N_2 + k_{31}N_3 + K_a \\
dN_2/dt &= -(k_{21} + k_{23})N_2 + k_{12}N_1 + k_{32}N_3 \\
dN_3/dt &= -(k_{31} + k_{32} + k_e)N_3 + k_{13}N_1 + k_{23}N_2
\end{aligned}
\tag{12.7}
$$

in which the k_{ij} are the transfer rate constants for substance N, K_a is the rate of intake of N (from an indefinite large reservoir), k_e is the excretion rate constant (from compartment 3), and N_i is the concentration of substance N in compartment i. These compartments can be defined in any way convenient for the problem at hand. They need not to be actual reservoirs; they can also be states of the substance. For example, to examine iodine metabolism one can define one compartment for the thyroid, another for free or inorganic bound iodine, and still another for protein-bound iodine. By using iterative curve-fitting techniques one can compare calculations using the compartment model with measurements of the disappearance rate of radioactive iodine from the thyroid, the rate of excretion of radioactive iodine, and the time-dependent concentration of radioactive iodine bound to proteins in the bloodstream.

12.4 Cybernetics

Cybernetics and information theory
A large class of mathematical models can be described by the branch of science called *cybernetics*. Cybernetics is defined by Wiener (1961) as "the science

285

of control and communication in the animal and in the machine". The term originates from the Greek word κυβερνητηζ, which means steersmanship. Cybernetics, indeed, treats behavior, rather than the things that behave. Its terminology would contain statements such as "this variable is undergoing an harmonic oscillation" without being concerned with whether the variable is actually a point of a turning wheel or a potential in an electric circuit. Its power, therefore, is the fact that it establishes a common language, expressing the common aspects of such diverse subjects as a room thermostat and the central nervous system. For biology in particular, the development of cybernetics may have great significance because it provides a method for the scientific treatment of systems of which complexity is an essential part. Living organisms are complex and their complexity is essential to their function; we know now that living systems are so dynamic and interconnected that alteration of one factor immediately causes alterations in a great many others. With cybernetics it may be possible to study this complexity as a subject in its own right.

Cybernetics and brain function

Cybernetics deals with automatons, sometimes called "switching systems" or just "robots". A computer is an automaton. Concepts developed out of the abstract theory underlying computing have been widely used to investigate the interaction of animals and machines with their environment. In biology it is especially the study of the nervous system that has drawn heavily on such concepts. This is not too amazing since one can hardly escape noticing the analogy between pulse conducting neuronal networks and the circuitry of a computer. Already in the early 1940s the idea of the "formal neuron" emerged. Formal neurons, an abstraction of what was know at that time about neuronal operation, operated upon and emitted at specified times binary (all-or-none) signals. Their junctions were either excitatory or completely inhibitory. They functioned by summing up their binary inputs algebraically and subsequently giving an output signal if, and only if, this sum exceeded a certain specified threshold. Networks made up of formal neurons could represent complicated logical formulas and, thus, symbolize the process by which the brain responds to and gives off stimuli. Moreover, a formal neuronal network provided with "receptors" and "effectors" is equivalent to an abstract representation of a computing system. Analogies like these, computing systems and formal neuronal networks on one hand and brain function on the other, have originally been developed by Wiener (1961) and are now used in the recent development of computer science that one calls artificial intelligence.

Genetic information and brain structure

The analogy between formal neuronal networks and brain function implies that real neuronal networks in a brain are circuits assembled according to "wiring diagrams". This conclusion has some interesting consequences, among which is that the circuitry in a brain, involving some 10^{10} neurons, represents in its structure an amount of information that far exceeds the genetic

286

information of a genome (the total of genetic information contained in the triplets of the nucleotide sequence of DNA). One could conclude then that brain structure, at least as far as its "wiring diagrams" are concerned, is only in part determined genetically. However, one has to be careful in using the term information in this context. H. Bremerman (1967) made this calculation, comparing the number of triplet combinations in a typical genome (4×10^9 nucleotide pairs) with the number of ways in which n neurons with m dentrites can make connections with each other. By putting $n = 10^{10}$ and $m = 10^2$ and introducing the constraint that each neuron can only make connections with a subset $n'(=10^4)$ of neurons (arising from the limiting length of dendrites), he arrives at slightly less than 10^{13} bits of information in brain structure, which compares to about 10^{10} bits of information in the genome (the number of bits of information in this case is equal to the logarithm to the base 2 of the number of all possible combinations). In this theory, however, the fact that the information content in bits of brain structure cannot be compared fully with the content of genome information in bits based on base sequences alone is completely ignored. In addition to this base sequence-determined genetic information there is also something which one could call *epigenetic* information, more subtle and harder to quantitatively determine. The meaning of epigenetic information becomes clear when we look at protein synthesis. We have seen in Chapter 4 that, for instance enzymes catalyze and control biological reactions by their tertiary, and often quaternary structure. Enzymes loose all their activity when by mild heating, or by changing the pH, quaternary and tertiary structure are destroyed; activity is recovered when the original conditions are restored. But genetic information (the base sequences in DNA) determines only the primary structure (the sequence of amino acids) of the enzymes. Enzyme function (bound to higher-order structure) is not genetically but epigenetically determined. It is important to note, however, that under the proper conditions the active higher-order structures are fully determined by the sequence of amino acids and, thus, ultimately by the information from DNA. Epigenetic information does not, therefore, come from sources other than DNA. It could be seen as a manifestation of the fact that the whole is often more than an assembly of the parts.

The neuronal code
It may be too simplistic to state, *a priori*, that the "wiring diagrams" of neuronal networks in a brain are determined the same way as the higher-order structures of enzymes are. In a sense a "wiring diagram" is comparable with the higher-order struture of an allosteric enzyme (its "structure" serves a specific function). It is, however, a far more complex system. A specific "wiring" of a neuronal network should serve to process a specific type of information. However, ablation studies on the brains of animals have shown that signal processing, such as visual integration, is not dependent on the specific details of the "wiring" in the visual cortex; even extensive destruction did not result in marked disintegration of function. A number of cybernetic models (based on formal neuronal networks) have been proposed and elaborated to cope with this

problem. One of these models assumes very complicated and extended networks in which incoming signals are coded in so-called error-insensitive codes (codes with a sufficient amount of redundancy, such that the "message" can be recognized even after extensive mutilation of the code).

The difficulty in applying cybernetic concepts to real neuronal systems is that the functional units are never identified by directly observing neurons and their interconnections. The successful application of cybernetics requires a precise knowledge of the *ensemble* of possibilities upon which the real system operates; in other words, what one needs to know is the code (or codes) of the nervous system. Thus, any real application of the concepts of cybernetics can only follow from a knowledge of what the neurons do and how they do it. This information can only be obtained from measurements of the firing patterns of neurons, the changing potentials of the electroencephalogram (EEG), and the generator potentials from the sensory systems.

12.5 Generalizations in biology

Generalizations
We conclude this chapter, and this book, by a look at the validity of the generalizations made by the biophysical approach described in this and in the preceding chapters. This approach is one among many possible approaches to biology. It differs from many others in that it is a predominantly molecular approach. We looked at the biological macromolecules and concluded that their structure has a pure physical basis. We looked at their function and found that their function is a logical consequence of their structure and is, thus, based on pure physical or physicochemical principles as well. Furthermore, we found, that to describe the way in which an enzyme catalyzes and regulates a metabolic process we need not specify what metabolic process really was catalyzed and regulated. Neither did we need to know in what cell of what organ in what organism this process took place. We found also that the enzyme came about by processes, involving nucleic acids and a protein synthesizing apparatus, which are exactly similar in all known forms of life. Finally, we found that these processes follow from and are based on well-known physicochemical principles. On a molecular basis then, generalization seems to be possible and quite legitimate.

The principles of biology
What do these generalizations tell us of the fundamental principles of biology? Are these fundamentals nothing other than the physics and chemistry of these structures, which arose out of inanimate matter by a process that "captures" a rare fluctuation away from the average evolution toward disorder and randomness and which, subsequently, evolved necessarily and deterministically to the selfpreserving and selfreproducing entities we know as living systems? Or is it that living systems are essentially different and that there is, perhaps,

288

another as yet unknown physical principle, a complementary principle, that would generalize physics to include these strange entities?

This latter opinion is held by the theoretical physicist W. Elsasser (1966). His argument for the singularity of living systems is that they are members of a class which is radically different from those classes formed by physical systems. This is a consequence of their complexity. Because of the complexity, of structures as well as processes, a living system is a rare realization out of an immense number of possible realizations (possible states) of matter in this kind of organization. All these possible states are determined according to pure physical principles, but the fact that (because of this immense number of possibilities) living system are membes of a strictly heterogeneous class, predictions based on sampling of the class is fundamentally impossible. Physical systems, on the other hand are members of homogeneous classes in which one member is indistinguishable from another. Sampling, and, therefore, statistical predictions are quite possible.

The idea that the progress of knowledge about the properties of matter is tied to a "loss of explanation" was originally proposed by N. Bohr (1933) who applied it to quantum mechanics. Elsasser asserts, with reference to the abstraction of the inhomogeneous classes, that uncertainties analogous to the quantum mechanical uncertainty are basic in biology. In this he sees an important tool for the design of a new theoretical biology.

These ideas seem to be somewhat contradictory to the universality of many processes in molecular biology. Moreover, many physical systems do belong to classes which satisfy Elsasser's definition of an inhomogeneous class quite well. What could be stated is, that the complexity of living systems makes an *a priori* prediction impossible but does not at all put into question an *a posteriori* explanation based on pure physical principles. But this is not a prerogative of living systems only; the shape and look of every rock picked up from a field has this property.

The frontiers

Of course, the last word has not yet been spoken in biology. There are still frontiers of knowledge beyond which properties may be revealed that may of may not radically change our views. Two of these frontiers seem to be the beginning stage and the present end stage of evolution. At the beginning stage there is the origin of the self-reproducing self-preserving system which we call a living organism. The theory of evolution is no longer a hypothesis full of speculations. Many aspects of it are amenable to experimentation and much is revealed in that way. There still is, however, the nagging problem of the evolution of the genetic code and the protein synthesizing apparatus. The latter consists of a number of macromolecular components which are all coded by DNA. Thus, the code can be translated only by translation products of the code. How and when has this closed system developed? How did the code develop and why is it universal? Is there a direct relation between the development of the code, its universality, and the evolution of the genetic apparatus?

289

At the other end is the central nervous system. Although, as has been shown in previous chapters, we are beginning to understand how a single neuron fires its spikes, we still have not the slightest idea of how this firing relates to the operation of even a simple neuronal network. Even the structure of relatively simple nervous systems, let alone that of brains, is unknown and as yet inaccessible to physiological experimentation. Thus, we are still far from having a notion about the operation of processes like perception, learning, and memory.

Finally, we are still far away from explaining life as a phenomenon in itself and in its relation to the cosmos. But new insights which came from the recent developments in the physics of non-stable systems far from thermodynamic equilibrium may show the directions in which the answer to these questions pehaps can be found. Such frontiers are challenging, especially for those who, led by previous successes, want to approach biology with the conceptuality of the physical sciences.

Bibliography

Ashby, W.R. (1956). "An Introduction to Cybernetics". Chapman and Hall, Ltd., London.
Bremermann, H. (1967). Quantitative Aspects of Goal Seeking, Self-Organizing Systems, in: "Progress in Theoretical Biology, Vol. 1" (F.M. Snell, Ed.), pp. 59–77, Acad. Press, New York.
Elsasser, W.M. (1959). *Op. cit.* (Chapter 1).
Elsasser, W.M. (1966). "Atom and Organism". Princeton Univ. Press, Princeton.
Fröhlich, H. (1969). Quantum Mechanical Concepts in Biology, in "Theoretical Physics and Biology" (M. Marois, Ed.), pp. 13–22, North Holland Publ. Cie, Wiley Intersci., New York.
Katchalsky, A., and Curran, P.F. (1967). "Nonequilibrium Thermodynamics in Biophysics". Harvard Univ. Press, Cambridge, Ms.
Kornacker, K. (1972). Living Aggregates of Nonliving Parts: A Generalized Statistical Mechanical Theory, in "Progress in Theoretical Biology, Vol. 2" (R. Rosen and F.M. Snell, Eds.), pp. 1–22, Acad. Press, New York.
Monod, J. (1970), "Le Hasard et la Nécessité". Ed. du Seuill, Paris.
Oosawa, F., and Higashi, S. (1967). Statistical Thermodynamics of Polymerization and Polymorphism of Protein, in: "Progress in Theoretical Biology, Vol. 1". (F.M. Snell, Ed.), pp. 80–164, Acad. Press, New York.
Prigogine, I. (1962). "Introduction to Nonequilibrium Thermodynamics". Wiley Intersci., New York.
Prigogine, I. (1969). Structure, Dissipation and Life, in: "Theoretical Physics and Biology" (M.. Marois, Ed.), pp. 23–52, North Holland Publ. Cie, Wiley Intersci., New York.
Prigogine, I. (1978). Time, Structure and Fluctuations, *Science* **201**, 777–785.
Rosen, R. (Ed.) (1985), Theoretical Biology and Complexity, Acad. Press, Orlando.
Wiener, N. (1961). "Cybernetics" (2nd edition), MIT Press, Cambridge, Ms.

Appendix I. Some elements of quantum mechanics

AI.1 The principle of quantization and the uncertainty principle

Quantization

The introduction of quantum mechanics at the beginning of the twentieth century was a major development of physics and, for that matter, science in general. Before that, physical theories seemed to be settled and in a rather satisfactory state. Mechanics obeyed the Newtonian laws, optical phenomena were explicable by the wave theory developed by Huygens, Young, Fresnel, and Hertz, and electricity and magnetism were ruled by the laws of Maxwell. There were, however, a few unsolved problems. One of these was the fact that the theoretical derivation of the density of black-body radiation as a function of its wavelength did not agree with the experimental determination of this quantity. In order to overcome this difficulty, Max Planck proposed (in 1900) that energy is not transferred continuously but in integral multiples of a fundamental amount called a quantum. If v is the frequency of the radiated energy, this quantum of energy is

$$E = hv \qquad \text{(AI.1)}$$

in which h is a universal constant known as Planck's constant.

It soon became apparent that this principle of quantization was indeed a universal one. Difficulties in the reconciliation of experimental data with theory could now be solved. An elegant example is the theory of specific heat which Debye developed by the application of Planck's ideas. Another is Bohr's explanation of the typical atomic line spectra, such as the Balmer lines in the hydrogen spectrum, by quantization of the energy states in atoms.

Einstein extended Planck's theory by proposing that electromagnetic radiation actually consists of little parcels of energy hv, called photons, which travel in a vacuum with velocity c. This assumption explained the photoelectric effect and the absorption and emission of electromagnetic radiation. A particulate description of light was proposed a long time before by Newton, but it was discarded in view of the overwhelming evidence for the electromagnetic wave theory. The new element brought in by Einstein was not so much a revival of Newton's photon theory but rather a recognition that electromagnetic radiation sometimes appears as a wave phenomenon and at other times as a stream of particles. These two aspects of light do not, as may appear at first sight, *exclude* each other as a description, but rather *complement* each other.

291

The French physicist Louis de Broglie showed that particles, such as electrons, protons, or neutrons also have this dualistic character. A wave is associated with the movement of particles which has a wavelength

$$\lambda = h/p \tag{AI.2}$$

in which p is the linear momentum of the particles and h, again, is Planck's constant. These "matter waves" or "de Broglie waves" can be demonstrated in phenomena such as electron diffraction.

The wave equation
The theoretical justification of this duality is *quantum mechanics*, a theoretical concept set up by Erwin Schrödinger and Werner Heisenberg which links the continuous characteristics of a wave with the discrete characteristics of a particle. This is done by replacing the equation of motion of a system by a wave equation, the so-called *Schrödinger wave equation*. The solution of this wave equation is a function of space and time, the *wavefunction* (the mathematical expression of quantum mechanics). This function represents the *state* of a system (for instance, an electron in an atom or a molecule); physical quantities such as energy, angular momentum, etc., can be calculated relatively easily once the wavefunction is known.

The Schrödinger wave equation is a differential equation generated by replacing a physical quantity β by a mathematical operator $\boldsymbol{\beta}$ and having the operator act on a function ψ in equations of the type

$$\boldsymbol{\beta}\psi = \beta\psi \tag{AI.3}$$

A mathematical operator is a *prescription* for one or a combination of mathematical operations, such as multiplication, division, and differentiation. In AI.3 it can mean, for example: "Differentiate the function ψ with respect to the space coordinates and multiply it with a number of constants." In AI.3 the operation is put into an equation; usually, solutions for such equations are possible only for a number of discrete values of β, the so-called *eigenvalues* (from the German word *eigen* which means "own" or "belonging") of the equation. The solutions belonging to the eigenvalues are called *eigenfunctions*.

If the operator in AI.3 is derived from the total energy of the system, it is called the *Hamiltonian* **H** of the system. The Schrödinger wave equation then becomes

$$\mathbf{H}\psi = E\psi. \tag{AI.4}$$

The eigenvalues E of this equation are the values of the energy in each of the states determined by the eigenfunctions ψ, the so-called wavefunctions. This operation turns the wave equation AI.4 into an equation which is analogous to the equation of motion of, for instance, a vibrating string.

The solution of the motion equation of a vibrating string, when solved for a stationary (time-independent) case, is the *amplitude* of a standing wave. The square of the amplitude is the *intensity* of the wave. In analogy, the solution of wave equation AI.4, for example, for a moving particle, is a wave-

292

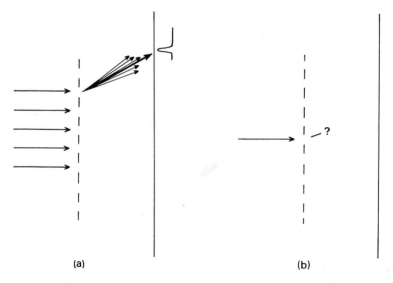

Fig. AI.1. A diffraction experiment (*a*) with a plane wavefront and (*b*) with a single photon.

function ψ, the square of which (rather of its absolute value) is the "intensity" of the particle at a given point. We can illustrate this by the following consideration: in diffraction experiments light behaves as a wave. In a typical experiment, such as the one illustrated in Fig. AI.1a, light from a point or line source passes through a slit and a diffraction pattern is recorded on a photographic plate. The process by which the photographic plate is sensitized can be explained, much like the photoelectric effect, by considering light as moving particles. From this point of view, then, the experiment can be considered as the passage of a stream of photons from the source to the plate. If it were possible to perform the experiment with one single photon (Fig. AI.1b), we would never obtain a diffraction pattern: at most one grain of the plate would be sensitized and we would not be able to predict the position of that grain in the photographic plate. We could, however, conduct such a single photon experiment many times, thus obtaining the equivalent of an experiment with many photons at one time. The diffraction pattern, thus, is an expression of the *probability* that a photon emitted from the source will strike the photographic plate at a given point. The diffraction pattern gives the intensity of the diffracted wave and the intensity is the square of the amplitude of the wave. So the square of the wavefunction (or rather of its absolute value) is then a measure of the probability of finding a particle in a given place. We could have followed exactly the same reasoning had we considered a stream of particles (electrons for instance) passing through a narrow slit. Diffraction patterns can actually be obtained from such a stream of electrons if the "slit" is small enough, which is the case for the openings between the molecules in a crystal.

In experiments of this sort, with light as well as with particles, the waves themselves are not observed. In fact, we never observe light as a wave but

293

rather as a quantized phenomenon, whether we detect the light with a photographic plate, a photocell, or our own eyes. We might, therefore, conclude (although this statement is inaccurate) that light "really" is a stream of photons and that the "waves" are the mathematical expressions of the way a stream of particles, such as electrons, moves. The similarity of such expressions with the expressions used for describing the motions in a vibrating string does not mean that light or a stream of electrons are vibrations of some mysterious medium, as "ether", or that the "wave" aspect and the "particulate" aspect are contradictory.

The uncertainty principle

An important corollary of quantum mechanics is the *uncertainty principle* of Heisenberg. This principle states that there are various pairs of variables (called canonically conjugated variables) which cannot be known with unlimited accuracy at the same time. This is not a technical but a fundamental limitation. Momentum and position are canonically conjugated variables; if the minimal uncertainty in the momentum is Δp and the minimal uncertainty in the position is Δx, their product is, according to Heisenberg's principle, a finite value equal to Planck's constant

$$\Delta x . \Delta p = h \qquad (AI.5)$$

(in many quantum mechanical calculations a polar coordinates, rather than Cartesian coordinates are used. The uncertainty then is $h/2\pi$). In "classical" mechanics it was assumed that it was possible to simultaneously determine these variables to any degree of accuracy. The uncertainty principle now states that this specification cannot be carried out beyond a certain limit. Suppose we set out to measure the position and the momentum of an electron in

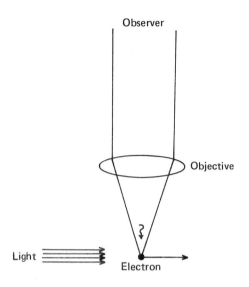

Fig. AI.2. An experimental setup to measure the position and momentum of an electron.

294

order to determine its motion. We can, in principle, observe the electron by a setup similar to that illustrated in Fig. AI.2. In such a setup we "illuminate" the electron with "light" and let the "light" scatter into our detection instrument. It is obvious that the shorter the wavelength of this "light", the more accurately we can observe the electron in a certain position. However, "light" quanta, now considered as particles with a momentum, have more energy and thus more momentum with shorter wavelength. Part of this energy is taken over by the electron after the collision, thus giving it an uncertainty in its momentum which can be calculated from the laws of conservation of energy and momentum (this is an effect which can be demonstrated experimentally and is known as the Compton effect). The attempt to gain accuracy in position by using "light" of shorter wavelength is, therefore, defeated by the loss of accuracy with the determination of momentum. A mathematical elaboration of this effect leads to relation AI.5.

The "wave character" of a stream of particles is complimentary to its "particle character", If we can design an experiment to determine the position of a particle, its momentum escapes our observation. If we want its momentum, we can only give a probability distribution (given by the wave function) for its position. This is the reason why the Bohr-Rutherford picture of an atom in which an electron revolves around the nucleus like a planet around the sun is incorrect. In a stable atom the energy of the electron, and its momentum, are determined. The electron's position with respect to the nucleus, therefore, cannot be determined. The best we can do is to state that the wave function, which in the stationary case is time-independent, gives us the probability distribution of the electron's position. The electrons, therefore, are often described as being "smeared out" in moving "clouds". The higher the probability in a certain area, the denser the cloud in that area.

Bibliography

Recommended introductory text in quantum mechanics: Wichmann, E.H. (1967, 1971). "Quantum Physics", Berkeley Physics Course, Vol. 4, McGraw Hill, New York.

Appendix II. Elements of equilibrium thermodynamics

AII.1 Definitions

Thermodynamics deals with the interconversions of various forms of energy and work. It does so by describing a *thermodynamic system* (which is a part of the universe separated from the rest) in terms of concepts and laws derived from the study of *macroscopic* phenomena such as pressure, volume, temperature, concentrations, etc. Any physical or chemical system can be described as a thermodynamic system or, simply, as a system.

A system is *isolated* when it can exchange neither energy nor matter with its surroundings; it is a *closed* system when it can exchange energy but cannot exchange matter with its surroundings; and it is an *open* system when it can exchange energy as well as matter with its surroundings. *Classical* or *equilibrium* thermodynamics describes systems which are in equilibrium or are undergoing *reversible* processes. A system is in equilibrium when it shows no internal tendency to change its properties with time. A system is in equilibrium when its *internal* parameters are completely determined by its *external* parameters. For example, consider a gas in a cylinder closed by a frictionless piston. This system is in equilibrium when the *external* pressure exerted by the piston on the gas is equal to the *internal* pressure of the gas, anywhere in the cylinder. If the gas is not in equilibrium we can still define the external pressure exerted by the piston, but this clearly is not a property of the system (the gas) itself; the internal pressure may vary from point to point in the gas before equilibrium is reached and its state cannot be determined by one value of the pressure equal to the external pressure of the piston. A system undergoes a reversible process when the path followed by such a process is one connecting intermediate states of equilibrium. For example, to reversibly expand the gas in a cylinder we have to release the pressure on the piston very slowly, in the limit infinitely slowly, so that at any instant the pressure of the piston equals the pressure anywhere in the gas. If the piston is drawn back suddenly, the expansion of the gas occurs irreversibly. In that case the intermediate states of the gas are no longer equilibrium states.

Classical thermodynamics is a harmonious and selfconsistent theory describing relations and correlations between the various parameters of a system in equilibrium or undergoing reversible processes. For irreversible processes, the laws of classical thermodynamics provide for a set of inequalities which only describe the *direction* of a change.

Reversible processes never occur in reality, since for the changes to be reversible they must occur at an infinitely slow rate. Equilibrium thermodynamics, therefore, does not seem adequate for real processes. Indeed, time does not come into the formalism equilibrium thermodynamics at all. In many cases, however, states which in reality were reached by processes moving along irreversible pathways can be realized, at least in "thought experiments", by reversible pathways as well. Thermodynamic quantities could thus be determined and this has in fact, proven to be very fruitful for the investigation of the properties and behavior of systems, including living ones.

The application of equilibrium thermodynamics to the description of the *course* of processes, especially of those in living systems, however, is limited. As we have mentioned before, equilibrium thermodynamics can only describe the direction of the changes and is unable to provide generalized quantative relations. *Thermodynamics of irreversible processes* or *nonequilibrium thermodynamics* may make possible the application of the formulations of thermodynamics to quantitative investigations of the processes of living systems. This is particularly important for the description of transport processes such as those occurring through biological membranes (see, for instance, Katchalsky and Curran, 1967).

AII.2 First and second laws of thermodynamics

If an *adiabatic* system (that is a system which cannot exchange energy or heat with its surroundings) is subject to external forces, work is performed. This work can be positive when it is done by the system *on* the surroundings, or negative when it is done on the system *by* the surroundings. If, for instance, the external force is a constant pressure P which causes a change in volume dV, the work is $dW = PdV$. Many experiments, particularly those of Joule, have confirmed that this amount of work, which changes the state of the system, is independent of the physical way in which the new state is reached. Thus, the compression work

$$W_{1\to2} = \int_1^2 P\, dV \qquad (AII.1)$$

is independent of the values assumed by P during the change and is only and fully determined by the initial and final states of the system. There must, therefore, exist an energy function (a *state function*) whose decrease represents the amount of work,

$$-dU = dW \qquad (AII.2)$$

and whose change is dependent only on the initial and final states of the system

$$W_{1\to2} = -\int_1^2 dU = U_2 - U_1 \qquad (AII.3)$$

298

Eq. AII.3 implies that for a cyclic process (when the system goes from state 1 tot state 2 and then back to state 1), the total change in U is zero:

$$\oint dU = \int_1^2 dU + \int_2^1 dU = U_2 - U_1 + U_1 - U_2 = 0 \qquad (AII.4)$$

in which the circular integral indicates integration over a cyclic process. Relations AII.3 and AII.4 apply to all thermodynamic state functions whose total change depends only on the values in the initial and final states of the system, regardless of the way in which the change is accomplished. The function U is the *internal energy* of the system. When the adiabatic restriction is lifted; the system can take up energy (heat) from its surroundings. This energy dQ is used to increase the internal energy and to do work:

$$dQ = dU + dW$$

or in a more familiar arrangement

$$dU = dQ - dW. \qquad (AII.5)$$

Relation AII.5 is the mathematical statement of the *first law of thermodynamics*. It states that energy can be neither created nor destroyed and thus expresses the impossibility of a *perpetuum mobile* (perpetual motion) of the first kind.

The first law provides an energy balance equation but it does not say anything about the direction in which natural processes spontaneously move. If two blocks of metal, one hot and the other cold, are placed in thermal contact with each other and isolated from their surroundings, we know by experience that the hot block becomes colder while the cold block becomes hotter; or, in other words, that heat flows from the hot block to the cold blok. This process continues until both blocks have the same temperature, somewhere between the two initial extreme temperatures. After they have reached this equilibrium temperature, no more heat flow takes place. We also know by experience that once in this state one of the blocks will never become hotter while the other becomes colder in a spontaneous way, although this would not at all violate the first law. Natural processes seem to proceed in a undirectional way, and the direction seems to be the one in which driving power is dissipated. A mathematical pendulum may swing periodically forever but an actual pendulum will stop after a while due to friction in its bearings and frictional resistance of the air, and it will never thereafter start swinging by itself. The *second law of thermodynamics* makes it possible to predict the direction in which natural processes occur. It tells us that for a natural process to occur, energy must be *dissipated* or *degraded*.

The second law can be stated verbally in many equivalent ways. A useful one for our purpose is the one given by W. Thomson (Lord Kelvin) in 1853: "It is impossible by means of inanimate material agency to derive mechanical effect from any portion of matter by cooling it below the temperature of the coldest of the surrounding objects". This quotation states the impossibility of a *perpetuum mobile* of the second kind: it is impossible that the cooling of a single body, the ocean for example, can provide useful energy, for example, energy required to drive a ship, in a cyclic manner, even though the removal of thermal energy from the water and its cyclic transformation into mechanical work would not violate the first law.

Thus, no positive work can be obtained from a cyclic process when it occurs at a constant temperature:

$$\oint (dW)_T \leqslant 0 \qquad (AII.6)$$

in which the subscript T means that the temperature is kept constant throughout the cycle. If a system produces a certain amount of work $W_{1\to 2}$ when it changes from a state 1 to a state 2 at a constant temperature, the amount of work $W_{2\to 1}$ required to bring the system back from state 2 to state 1 is larger or equal to $W_{1\to 2}$:

$$W_{1\to 2} \leqslant W_{2\to 1}. \qquad (AII.7)$$

The equalities of both AII.6 and AII.7 apply when the process occurs *reversibly*. In that case one can prove that

$$\oint \frac{dQ}{T} = 0 \qquad (AII.8)$$

which means that when the system changes reversibly from state 1 to state 2 the value of the integral

$$\int_1^2 \frac{dQ}{T}$$

depends only on the initial state 1 and the final state 2 and not at all on the way in which the change has taken place. T stands for the absolute temperature. For reversible processes, therefore, we can define a function S such that its change dS is

$$dS = \frac{dQ}{T}. \qquad (AII.9)$$

According to AII.8, for a reversible process, $dS = 0$ and, when the system changes from a state 1 to a state 2,

$$\int_1^2 dS = S_2 - S_1 \qquad (AII.10)$$

thus indicating that the change in the function S when the system moves from state 1 to state 2 depends only on the initial and the final values of the function. The state function S is called the *entropy*. Its definition as described above was introduced by R. Clausius in 1865.

Although entropy is defined above only for reversible processes, the fact that the entropy difference between two states of a system depends only on its initial and final value (Eq. AII.10) also makes it a useful function for natural irreversible processes. The change in entropy in going from a state 1 to a state 2 is always the same, that is, irrespective of the path between 1

300

and 2 and of whether or not the process is reversible. In order to investigate the change in entropy resulting from a natural irreversible transition from a state 1 to a state 2 by an isolated system, we can think of the system returning to its initial state 1 along a reversible path. We then can apply the definition for the entropy of reversible processes and arrive at an expression for the entropy difference between state 1 and state 2. If this is done we will discover that the second law requires that, for an irreversible process in an isolated system, the entropy always increases:

$$\Delta S = S_2 - S_1 > 0. \tag{AII.11}$$

Since all naturally occurring processes are irreversible, any change that actually occurs spontaneously in an isolated system is accompanied by a net increase in entropy. Stated the other way around, but equally valid, if there is any conceivable process for which the entropy can increase, it will occur spontaneously. For irreversible processes inequality AII.11 expresses the second law, indicating the direction in which spontaneous processes will proceed. This direction is always towards a maximum entropy. When this maximum is reached no change will occur spontaneously and the system is in equilibrium.

At equilibrium, relation AII.9 is valid. This relation represents the second law for reversible processes. The introduction of this equation into Eq. AII.5 gives the expression for the combined first and second laws

$$dU = T\,dS - dW. \tag{AII.12}$$

From this we derive that for an adiabatic process

$$dU = -dW \tag{AII.13}$$

since $dQ = 0$. For a reversible isothermic cycle Eq. AII.12 becomes

$$\oint dU = T \oint dS - \oint dW \tag{AII.14}$$

and, since both dU and dS depend only on final and initial states and therefore have zero circular integrals,

$$dW = 0 \tag{AII.15}$$

which mathematically expresses Lord Kelvin's formulation of the second law for an isothermic reversible cycle.

AII.3 Entropy

The physical meaning of entropy is not intuitively clear from Clausius' definition. It can be seen as that part of the heat-energy term that must be transported across a temperature difference in order to produce work. The entropy is an *extensive* quantity which has to be transported across a potential difference. The temperature is an *intensive* (potential) quantity which provides such a potential difference. All the energy terms, except the internal energy,

301

consist of an extensive and an intensive (potential) part. Work is produced when there exists a difference of potential across which the conjugated extensive quantity can be transported. The work function, for instance, can contain many terms in addition to expansion work:

$$dW = PdV - fdl - \psi de - \Sigma_i \mu_i dn_i + \ldots \quad (AII.16)$$

in which changes in volume dV, length dl, charge de, and the number of moles of a species i in a chemical reaction dn_i are extensive quantities which can produce work when transported over "potential" differences provided by pressure P, mechanical force f, electrical potential ψ, and chemical potential μ_i, respectively. Introducing AII.16 in AII.12 we obtain the equation of Gibbs

$$dU = T\,dS - PdV + fdl + \psi de + \Sigma_i \mu_i dn_i + \ldots \quad (AII.17)$$

which takes into account all possible changes in extensive properties (dS, dV, dl, de, dn_i) and relates the total change in internal energy to the sum of the product of intensive (potential) quantities (T, P, f, ψ, μ_i) and the changes in the extensive properties.

With the development of statistical mechanics, the concept of entropy became more tangible. Boltzmann has shown that entropy is directly related to the number of configurations in which a system can be realized. Consider, for example, a situation in which a large number (N) of apples has to be distributed over two baskets. Obviously, the most unlikely distribution is that in which there are no apples in basket 1 and N apples in basket 2. This distribution can be accomplished in only one way; the system can be realized by only one configuration. A distribution in which one apple is in basket 1 and $N-1$ apples in basket 2 is a little more probable. This can be accomplished in N ways. The distribution in which 2 apples are in basket 1 and $N-2$ in basket 2 can be achieved in even more ways by adding to any of the N apples which can be placed in basket 1 any of the remaining $N-1$ apples, hence in $N(N-1)$ ways; this system, thus, can be realized by $N(N-1)$ configurations. By putting more and more apples in basket 1 the number of configurations in which the different distributions can be realized, and thus the probability or *randomness* of the distribution, will increase until each basket has $N/2$ apples.

If Ω is the number of configurations in which a system can be realized, the entropy is

$$S = k \ln \Omega \quad (AII.18)$$

in which k is the *Boltzmann constant*. The entropy is, thus, a measure of randomness, and to state that a spontaneous process always moves toward a maximum entropy is equivalent to saying that a process spontaneously moves toward a maximum randomness. The example of the apples in the two baskets is comparable to the two copper blocks of different temperature in thermal contact, or to a vessel in which a screen with a hole in it separates a zone containing a gas at high concentration from a zone containing gas at low concentration. In both cases a spontaneous process occurs from a state which is less probable, less random, to a state which is more probable, more random; heat flows from the hot block to the cold block until both blocks have the same intermediate temperature and gas molecules diffuse from the zone of high concentration to the zone of low concentration until the concentration throughout the vessel has the same intermediate value. In both cases, the process stops when the entropy has the maximum value of the equilibrium point.

AII.4 Thermodynamic potentials

Linear combinations of thermodynamic state functions, such as the internal energy U and the entropy S, are also state functions; they are called *thermodynamic potentials*. Sometimes such combinations are more amenable to experimental determination. An example is the enthalpy H, defined as

$$H = U + PV \qquad \text{(AII.19)}$$

from which it follows that

$$dH = dU + PdV + VdP. \qquad \text{(AII.20)}$$

This function is particularly useful in situations where the pressure remains constant. In that case

$$dH = dU + PdV. \qquad \text{(AII.21)}$$

If the only work done in a system is expansion work PdV (which is the case, for example, for a chemical reaction carried out in a test tube), application of the first law AII.3 yields

$$dH = dQ. \qquad \text{(AII.22)}$$

The enthalpy change dH, therefore, is the heat taken up (when positive) or given off (when negative) when a reaction occurs under isobaric (constant pressure) conditions and when no work other than expansion is done. It is often called the *reaction heat* or *heat content*. Many compounds have a characteristic reaction heat. In the case of combustion this heat is the heat of combustion when one gram molecule of the compound is completely burned in molecular oxygen. The heat of combustion is equal to the molar enthalpy change of the reaction. When heat is given off to the surroundings the enthalpy change is negative and the reaction is *exothermic*. When heat is taken up from the surroundings the enthalpy change is positive and the reaction is *endothermic*.

Another useful thermodynamic potential is the *Helmholtz free energy F*, defined by

$$F = U - TS \qquad \text{(AII.23)}$$

from which it follows that

$$dF = dU - TdS - SdT. \qquad \text{(AII.24)}$$

This potential function is useful for isothermal processes (processes occurring at a constant temperature) for which

$$dF = dU - TdS. \tag{AII.25}$$

The maximum work can be obtained from a process when it occurs reversibly. In that case we obtain, by introducing expression AII.12 into AII.25,

$$-dF = dW_{max}. \tag{AII.26}$$

The decrease in free energy, therefore, represents the maximum amount of work which can be performed under isothermal conditions. From Eqs. AII.26 and AII.23 we can deduce that the free energy is that part of the energy which, at a constant temperature, is useful for work. If a system is left alone, spontaneous processes which go in the direction of increasing entropy ($dS > 0$) also go in the direction of decreasing free energy. At equilibrium the entropy is at a maximum ($dS = 0$) and the free energy is at a minimum ($dF = 0$).

The Gibbs free energy is a potential function which is particularly useful for biochemical and biological processes, which usually occur not only at a constant temperature but also under constant pressure. The Gibbs free energy G is defined as

$$G = U - TS + PV. \tag{AII.27}$$

At constant temperature and pressure

$$dG = dU - TdS + PdV = dH - TdS. \tag{AII.28}$$

Using AII.12 we obtain, for reversible processes,

$$-dG = dW_{max} - PdV = dW_{net} \tag{AII.29}$$

in which dW_{net} is all of the work other than expansion or compression work. Willard Gibbs called dW_{net} the useful work. The decrease in Gibbs free energy, therefore, represents the maximum amount of useful work (other than expansion or compression work) which can be obtained from a process occurring at constant temperature and pressure. Of course, at equilibrium ΔG is also equal to zero.

AII.5 The chemical and the electrochemical potentials

When we consider biological systems we must take into account that they are open systems; matter is brought into or taken away from the system by both transport processes and chemical reactions. It is important, therefore, that we generalize our treatment in such a way that such open systems are included. This can be done by including the change in composition brought about by the chemical potential μ in the work function (Eq. AII.16).

The variation of the Gibbs free energy at constant temperature and pressure, dG, is given by Eq. AII.28. When we substitute the variation of the internal energy given by the Gibbs equation AII.17, thereby ignoring all contributions to the work function except expansion work and "chemical" work, we get

$$dG = SdT + VdP + \sum_i \mu_i dn_i. \qquad (AII.30)$$

The change in free energy by the addition (or removal) of component i to (or from) the system, keeping temperature, pressure, and the amount of all components other than component i constant, is given by

$$\frac{\partial G}{\partial n_i}\bigg|_{T,P,n_{j \neq i}} = \mu_i. \qquad (AII.31)$$

The chemical potential is, thus, defined as the increase of the Gibbs free energy on the reversible addition of component i, holding temperature, pressure, and the amount of all other components constant.

For a pure one-component system the chemical potential is simply the Gibbs free energy per mole of the component. Therefore, μ_i is sometimes called the *molar free energy* of i when the concentration of i is expressed in moles. For a multicomponent system it can be mathematically shown that the total free energy at a given temperature and pressure is given by

$$G = \sum_i n_i \mu_i. \qquad (AII.32)$$

For an ideal gas the chemical potential can be calculated per mole by using the ideal gas equation

$$PV = RT \qquad (AII.33)$$

in which R is the gas constant. Since on expansion the gas performs work given by

$$dW = PdV \qquad (AII.34)$$

the change in chemical potential must be

$$d\mu_g = -PdV. \qquad (AII.35)$$

When we differentiate AII.33, we obtain

$$d(PV) = PdV + VdP = d(RT) = 0 \qquad (AII.36)$$

at constant temperature. Hence,

$$PdV = -VdP \qquad (AII.37)$$

and

$$d\mu_g = VdP \qquad (AII.38)$$

at constant temperature. Solving for V from Eq. (AII.33) and substituting it in AII.38 gives

$$d\mu_g = RT(dP/P) \qquad (AII.39)$$

at constant temperature. Integration of Eq. AII.39 yields

$$\mu_g - \mu_g{}^\circ = RT \ln (P/P_0). \qquad (AII.40)$$

The chemical potential μ_g is equal to $\mu_g{}^\circ$ when the pressure of the gas is equal to P_0. We can use this as a standard against which the chemical potential can be measured by simply putting P_0 equal to 1 atm at any particular temperature. Eq. AII.40 then becomes

$$\mu_g = \mu_g{}^\circ + RT \ln P_g. \qquad (AII.41)$$

The calculation of the chemical potential of a solute in a solution proceeds analogously. We now use the Van 't Hoff equation

$$\pi_i = c_i RT \qquad (AII.42)$$

which is valid for ideal (infinitely diluted) solutions. π_i is the osmotic pressure and c_i is the concentration of solute i. The work per mole performed by the osmotic pressure is $\pi_i d(1/c_i)$, and thus

$$d\mu_i = -\pi_i d(1/c_i) = c_i RT \, d(1/c_i) = RT \, d(\ln c_i). \qquad (AII.43)$$

Integration yields

$$\mu_i - \mu_i{}^\circ = RT \ln c_i/c_0). \qquad (AII.44)$$

Again we can use a value $c_0 = 1$ mole to define the standard potential $\mu_i{}^\circ$, giving

$$\mu_i = \mu_i{}^\circ + RT \ln c_i. \qquad (AII.45)$$

We can apply the same procedure when calculating the chemical potential of a solvent. In a solution, the solvent also has a concentration; there is less solvent per unit volume in a solution than in the pure solvent. We prefer, however, to use the mole fraction instead of moles for the solvent concentration. The mole fraction is defined as

$$x_j = n_j / \sum_i n_i, \quad i = 1, \ldots, j, \ldots \qquad (AII.46)$$

The use of mole fraction for the solvent has the convenient property of being equal to 1 for the pure solvent. The chemical potential of the solvent in a solution is given by

$$\mu_s = \mu_s{}^\circ + RT \ln x_s \qquad (AII.47)$$

whereby $\mu_s = \mu_s{}^\circ$ for the pure solvent.

Eqs. AII.41, AII.45, and AII.47 are derived for ideal gases and ideal solutions. In such systems there are no molecular interactions: strictly speaking, the derived formulas are valid only for such systems. It is still formally correct, however, to use the equations, provided that we replace the gas pressure P_g in AII.41 by a quantity called *fugacity*, f_g, defined by

$$f_g = \alpha_g P_g \qquad (AII.48)$$

and the concentration c_i (or mole fraction x_s) in AII.45 and AII.47 by the *activity* a_i, defined by

$$a_i = \alpha_i c_i. \tag{AII.49}$$

The coefficients α_g and α_i are the *fugacity coefficient* and the *activity coefficient*, respectively. They are purely empirical coefficients. If the pressure or the concentration is sufficiently low (as in diluted systems) the use of pressure and concentration is a reasonable approximation.

If a solute bears an electric charge (for instance, an ionic species in a solution of an electrolyte) there is also an electrical potential that has to be taken into account. The ability to do useful work in that case is due to the chemical potential (related to concentration) *and* to the electrical potential (related to electric charge). For such a solute we define the *electrochemical potential*

$$\tilde{\mu}_k = \mu_k + z_k \mathscr{F} \psi = \mu_k^\circ + RT\,c_k + z_k \mathscr{F} \psi \tag{AII.50}$$

in which z_k is the valence of ionic species k, ψ is the electrical potential, and $\mathscr{F}$ is the *Faraday constant*, which is the electronic charge per mole (since the electronic charge $\epsilon = 1.6 \times 10^{-18}$ coulomb and Avogadro's number $A = 6.02 \times 10^{23}$, $\mathscr{F} = A\epsilon = 96,489$ coulomb per mole; in calories per volt, a unit more useful for our purpose, $\mathscr{F} = 23,061$, since 1 cal = 4.184 joules).

AII.6 The standard free energy of a chemical reaction

We can, in a fairly general way, represent a chemical reaction by the equation

$$aA + bB \rightleftharpoons pP + qQ \tag{AII.51}$$

in which, a, b, p, and q are the stoichiometric coefficients. For such a reaction the change of free energy is given by the difference between the sum of the chemical potentials of the products (the components on the right-hand side of the equation) and the sum of the chemical potentials of the reactants (the components on the left-hand side of the equation), each chemical potential being multiplied by the appropriate stoichiometric coefficient. Thus,

$$\begin{aligned}
\Delta G &= p\mu_Q + q\mu_Q - a\mu_A - b\mu_B \\
&= p\mu_P^\circ + pRT \ln c_P + q\mu_Q^\circ + qRT \ln c_Q \\
&\quad -a\mu_A^\circ - aRT \ln c_A - b\mu_B^\circ - bRT \ln c_B
\end{aligned} \tag{AII.52}$$

or

$$\Delta G = \Delta G^\circ + RT \ln (c_P{}^p c_Q{}^q / c_A{}^a c_B{}^b) \tag{AII.53}$$

when we lump the standard potentials μ_i° together in one symbol ΔG°.

At equilibrium $\Delta G = 0$. Hence, since ΔG° is a constant, the logarithm in Eq. AII.53 must be a constant at equilibrium. Thus,

$$\frac{c_P^p c_Q^q}{c_A^a c_B^b} = K \qquad \text{(AII.54)}$$

in which K is called the *equilibrium constant*. According to AII.53

$$\Delta G^\circ = -RT \ln K. \qquad \text{(AII.55)}$$

ΔG° is called the *standard free energy change* of the reaction and is a measure of the direction in which the reaction spontaneously proceeds toward equilibrium. It can be seen from AII.54 and AII.55 that ΔG° is negative when the concentrations of the products at equilibrium are higher than those of the reactants. A negative value of ΔG° means that the reaction proceeds spontaneously in the direction of the products. The reaction is said to be *exergonic*. A positive value of ΔG°, on the other hand, means that it will cost energy to push the reaction in the direction of the products. The reaction then is *endergonic*. The energy necessary to drive the reaction in the direction of the products can be obtained, for example, by coupling the reaction to an exergonic reaction which has a value of ΔG° at least as negative as the ΔG° of the original reaction is positive.

There are two important things to note about the standard free energy change. First, it is not an *actual* change in free energy; it is just a measure for the direction in which a reaction would go when it spontaneously proceeds to equilibrium and how far it would go. Second, its value is not a measure for the *rate* of the reaction. It is quite possible that a strongly exergonic reaction which has a large negative ΔG° proceeds with an infinite slow rate, for instance, because it has a high activation energy barrier. In many cases a catalyst, such as an enzyme in a biological system, can lower this barrier so that the reaction can proceed at a high rate toward equilibrium.

AII.7 Oxidation-reduction potentials

Oxidation-reduction reactions are reactions which involve the transport of electrons. A substance is said to be reduced when it receives electrons. A substance becomes oxidized when it loses electrons. In aqueous media (such as we usually have in living cells) a reduced substance can pick up protons from the medium so that reduction becomes equivalent to the addition of hydrogen. An oxidized substance can bind to oxygen because of the strong electronegativity of oxygen. Essentially, the reaction is the transport of electrons; often, it does not go together with protonation or oxygenation, as in the cytochrome reactions of the respiratory chain.

Since oxidation-reduction reactions amount to the exchange of negative electric charge, it is convenient to express the free energy change of such reactions in units of electric potential. This can be done as follows.

We can represent a reaction in which substance B reduces a substance A by

$$A_{ox} + B_{red} \rightleftharpoons B_{ox} + A_{red}. \qquad (AII.56)$$

This reaction can be split into two half-reactions

$$A_{ox} \rightleftharpoons A_{red} \qquad (AII.56a)$$

and

$$B_{red} \rightleftharpoons B_{ox}. \qquad (AII.56b)$$

The free energy change of the half-reaction AII.56b is the free energy change that is associated with the transition of B from the reduced to the oxidized state. According to AII.53, this is equal to

$$\Delta G = \Delta G^\circ + RT \ln \frac{B_{ox}}{B_{red}} = \Delta G^\circ + RT \ln \frac{\alpha}{1 - \alpha} \qquad (AII.57)$$

where α is the fraction of the amount of B which is oxidized. We can convert this to an electrical potential by dividing it by $-z\mathscr{F}$, in which z is the number of transported electrons in the reaction. (The negative sign comes in because it is a negative electric charge that is transported). Thus,

$$E_B = E_B{}^\circ - \frac{RT}{2\mathscr{F}} \ln \frac{\alpha}{1 - \alpha}. \qquad (AII.58)$$

The potential E is the *oxidation-reduction potential*, or redox potential, and can be measured against a standard. A convenient standard for this purpose is the *hydrogen electrode*. A hydrogen electrode is a platinum electrode immersed in an acid solution against which hydrogen gas is bubbled. The half-reaction at the electrode, occurring when the electrode is part of a closed circuit, is

$$H_2 \rightleftharpoons 2H^+ + 2e \qquad (AII.59)$$

and the potential is

$$E = E^\circ - \frac{RT}{2\mathscr{F}} \ln \frac{c_{H^+}^2}{c_{H_2}}. \qquad (AII.60)$$

When we use hydrogen gas at a pressure of 1 atm, $c_{H_2} = 1$ and AII.60 becomes

$$E = E^\circ - \frac{RT}{\mathscr{F}} \ln c_{H^+}. \qquad (AII.61)$$

Since $\ln c_{H^+} = 2.3 \ {}^{10}\log c_{H^+} = -2.3$ pH, T $= 298°$K ($= 25°$C, the standard temperature used for electrochemical cells), R $= 1.98$ cal/mole degree, and $\mathscr{F} = 23,061$ cal/V,

$$E = E^\circ + 0.06 \text{ pH}. \qquad (AII.62)$$

The potential E measured by the hydrogen electrode emersed in a 1 N acid solution (pH $= 0$) is chosen by convention to be zero. Hence, $E = E^\circ = 0$

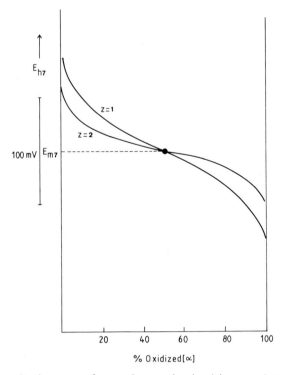

Fig. AII.1. Redox titration curves for a redox reaction involving one electron ($z = 1$) and a redox reaction involving two electrons ($z = 2$). E_{h7} is the redox potential standardized against the hydrogen electrode at pH = 7. E_{m7} is the midpoint potential.

at pH = 0. Against this standard all redox potentials can be measured in an absolute way.

In living cells the ambient pH is not far from neutral. Thus, it is more convenient to take the standard at pH = 7, rather than at pH = 0. Therefore, where biologically important redox couples are concerned, the potential of the hydrogen electrode at pH = 7 is chosen to be zero. Thus,

$$E_{h7} = 0 = E_{h7}° + 0.42 \qquad\qquad (AII.63)$$

(in which E_{h7} is the redox potential, measured with the restandardized electrode), and the standard value (the redox potential of hydrogen) $E_{h7}° = -0.42$ volt.

Using the hydrogen electrode at pH = 7 as a standard, we can rewrite Eq. AII.58 as follows:

$$E_{h7} = E_{m7} - \frac{RT}{2\mathscr{F}} \ln \frac{\alpha}{1 - \alpha}. \qquad\qquad (AII.64)$$

in which E_{m7} is the midpoint potential, measured when half of the substance is in the oxidized state and half of it is in the reduced state. The midpoint potential is characteristic for a redox couple.

310

In Fig. AII.1, E_{h7} is plotted as a function for α for two values of z. Such curves are called *redox titration curves*. They can be obtained by titrating a redox reaction (either in the direction of reduction or in the direction of oxidation) with some indicator (a spectral change, for example) and simultaneously measuring the potential. The inflection point is at the midpoint potential.

Bibliography

Recommended text in equilibrium thermodynamics:
Spanner, D.C. (1964). "Introduction to Thermodynamics". Acad. Press, New York).

Index